Food Engineering Series

For further volumes:
http://www.springer.com/series/5996

Tamara Allaf • Karim Allaf
Editors

Instant Controlled Pressure Drop (D.I.C.) in Food Processing

From Fundamental to Industrial Applications

Springer

Editors
Tamara Allaf
ABCAR-DIC Process
La Rochelle, France

Karim Allaf
University of La Rochelle
La Rochelle, France

ISSN 1571-0297
ISBN 978-1-4614-8668-8 ISBN 978-1-4614-8669-5 (eBook)
DOI 10.1007/978-1-4614-8669-5
Springer New York Heidelberg Dordrecht London

Library of Congress Control Number: 2013948231

Printed on acid-free paper

Springer is part of Springer Science+Business Media (www.springer.com)

Preface: A Question of Thermodynamics

It is a well-known fact that any thermodynamic system consists of a very large number of particles. . . so large that to count the molecules in a 1-g drop of water, at a rate of 100 molecules per second, you would need 10,000 billion years! A child celebrating his birthday and blowing out the candles at a rate of 1,000 molecules of air per second would take some 400 billion years to expel 0.5 L of air. . ..

These particles are in uninterrupted motion so the air molecules you have around you, in a "static" situation at about 18 °C, have an average *translational fluctuation velocity* of about 500 m/s. . . or 1,800 km/h!

Three degrees of freedom are associated with the translation movement of a particle in three-dimensional space.

The average density of the kinetic energy of translational fluctuation per mole can be adopted as the basis of the true thermodynamic temperature—let's call it θ! It is expressed in J/mol. While the conventional measure of temperature T, for historical reasons, was based on a sensorial approach and expressed in °C, °F, or K, the true thermodynamic temperature is a physical concept which correlates with T (in K) through $\theta = 3/2\ RT$. This is very important and it makes much more sense to let our students, researchers, and engineers think of temperature in terms of J/mol instead of °C. . . given that 12.5 J/mol represents approximately 1 K.

T (°C)	−273	−193	−113	−33	−18	0	20	37	42	100
T (K)	0	80	160	241	255	273	293	310	315	373
θ (J/mol)	0	1,000	2,000	3,000	3,182	3,406	3,656	3,868	3,930	4,652

This translational fluctuation velocity is not uniform for all particles and a "normal" distribution allows the highest translational fluctuation energy particles in a liquid to fly (evaporate) from the surface to the surrounding atmosphere. The higher the average translational fluctuation energy, the greater the number of particles flying off. In other words, the higher the temperature of a liquid, the higher the vapor pressure.

When the energy brought by "evaporation" (flying particles) is supplied by the system itself, the operation is called "autovaporization"; (It is like a family paying for a flight for one of its members; the flight can be financed using the family's own capital, it is self-financed).

The study of industrial unit operations inevitably requires a fundamental analysis of thermodynamic transformations, various mass and heat transfers, chemical and enzymatic reactions, mechanical/rheological behavior and characteristics. Studies have been carried out on a large number of transfer phenomena that occur between or inside media. Here, change over time reveals the exchange processes in quasi-static systems: the homogeneous isotropic translation movement of particles in different amounts and in different zones establishes gradients of temperature and concentration which act as driving forces.

Some operations involve *transitory periods* which are so short that they are generally considered to be negligible. However, the impact of such transitory periods has been proved to be so effective that it is essential that they would be the subject of specific studies. During the very short period when the pressure is dropped abruptly, the initial isotropic translational fluctuation velocity ceases being random in one of the three dimensions; then two dimensions of the initial three dimensional-translational fluctuation velocity define the temperature, whose value, from an asymptotic point of view, becomes $\theta_m = 2/3\ \theta_i$. This modification of temperature is achieved without an exchange of energy with the external medium and does not respect the conventional quasi-static transfer laws. When this can achieve autovaporization at the industrial level, the particular impact is to produce higher yields thanks to a fall in temperature, which widely overshoots the equilibrium level. This can be considered as the basis for improving industrial operation performances in terms of energy consumption and environment features.

Thermodynamics then has two asymptotic limits: the quasi-static field, which concerns long transformations; and the very short and specific instantaneous field. Instant controlled pressure drop (DIC) belongs to the latter field in one of the technological applications described in this book.

Other transitory-type processes have been considered as innovative approaches in the intensification of industrial unit operations. The main concerns in this field have been the possibility of using such transitory stages of thermodynamic transformations to establish much more effective mass transfers. Using the total pressure gradient as a driving force instead of the vapor pressure of a standard diffusion transfer (Fick-type transfer) allows us to define the instant autovaporization which can markedly improve drying and essential oil extraction.

Temperature, Definition, and Presentation

Temperature is now well defined: it is the density of the kinetic energy of the translational fluctuation motion of the particles in a system. Thus it is possible to define the term "temperature" as a physical quantity according to the relationship between the thermodynamic-temperature θ and the density of the kinetic energy per mole for the translational fluctuation of particles in a system:

$$\theta = \frac{3}{2}RT = \frac{1}{2}MV_t^2 \qquad (1)$$

where θ is the quantity reflecting the system temperature (J mol^{-1}); R is a constant called the universal ideal gas constant ($R = 8.314$ J mol^{-1} K^{-1}); T is the temperature expressed in Kelvin K; M is the molar mass of the particles forming the system being studied; V_t is the mean translational velocity of the particles; it is a random 3-dimensional isotropic velocity with a statistically homogeneous value $V_t^2 = V_x^2 + V_y^2 + V_z^2 = 3V_x^2$.

Whatever the size and the composition, a system is mainly characterized by an extremely large number of huge and randomly moving particles. The total velocity of the system derived from each particle can be defined as follows.

Total velocity:

$$\vec{v}_G = \frac{\sum m_i \vec{v}_i}{\sum m_i} = \frac{\sum m_i \vec{v}_i}{m} \qquad (2)$$

When the total velocity $\vec{v}_G$ is nil, the system is static, although $\vec{v}_i \neq 0$.

When the composition of these different velocities leads to a total velocity that is different from zero, the system has an apparent motion which has to be added to the internal fluctuating velocities.

Absolute and relative velocities:

$$\vec{v}_i = \vec{v}_G + \vec{V}_i \qquad (3)$$

where $\vec{v}_i$ is the absolute (laboratory frame) velocity of the particle i; $\vec{v}_G$ is the absolute (laboratory frame) total velocity (of the center of mass) of the system; $m = \sum m_i$ is the mass of the system, which is the sum of the masses of each particle; $\vec{V}_i$ is the relative (center of mass frame) velocity of the particle i. Its average value rapidly leads to a homogeneous distribution of particles and particle velocities in the system which becomes 3-D isotropic and statistically homogeneous.

Thus, the integral kinetic energy of the system is due to the two components of velocity: macroscopic (displacement of the center of mass, with the general external kinetic energy) and random (internal kinetic energy of fluctuation, not necessarily limited to the translational motion).

Integral kinetic energy E_{k} of a system:

$$E_{\mathrm{k}} = \sum \frac{1}{2} m_i v_i^2 = \frac{1}{2} m v_G^2 + \sum \frac{1}{2} m_i V_i^2 \tag{4}$$

The translational part of internal kinetic energy of fluctuation $\sum \frac{1}{2} m_i V_i^2$ is linked to 3 as degrees of freedom (DoF).

The translational part of the kinetic energy of fluctuation:

$$\sum_{j=1}^{3} \sum \frac{1}{2} m_i V_{\mathrm{t},j,i}^2 = \frac{1}{2} m V_{\mathrm{t}}^2 \tag{5}$$

Other possible motions (vibration, rotation) constitute a complementary degree of freedom. By assuming δ is the total degrees of freedom, the total internal kinetic energy of fluctuation is as follows.

Total internal kinetic energy of fluctuation:

$$E_{\mathrm{k,f}} = \sum \frac{1}{2} m_i V_i^2 = \delta \frac{1}{2} m V_{x1}^2 = \frac{\delta}{3} \left(\frac{1}{2} m V_t^2 \right) \tag{6}$$

The integral kinetic energy E_{k} is the sum of the total external kinetic energy $E_{\mathrm{k},G} = \frac{1}{2} m v_{\mathrm{CM}}^2$ and the total internal kinetic energy of fluctuation $\frac{\delta}{3} \left(\frac{1}{2} m V_{\mathrm{t}}^2 \right)$:

$$E_{\mathrm{k}} = E_{\mathrm{k},G} + E_{\mathrm{k,f}} = \frac{1}{2} m v_G^2 + \frac{\delta}{3} \left(\frac{1}{2} m V_{\mathrm{t}}^2 \right) \tag{7}$$

This is the situation of a static system (free of interference from the external environment) or a system with quasi-static transformations. The study of thermodynamic systems is an essential element in the analysis of physical, chemical, and biological phenomena.

Once it is isolated, or during quasi-static transformations, the system can have the density of the translational kinetic energy of fluctuation, which is usually three dimensional. It is hence possible to define the term "temperature" as a physical quantity according to the relationship shown in (1).

Before it was defined as a physical quantity, temperature was indicated using "thermometers." This is to link the temperature with physical impacts (dilation, thermoelectricity).

In the case of fluids, the large number of particles and the random translational velocity define the *pressure* as the parameter that is closely coupled to the rate of momentum transferred to the unit area of the wall that defines the system.

Thermodynamic Parameters

A thermodynamic system is defined by its internal energy U. This is the sum of the internal kinetic energy of fluctuation:

$$E_{\mathrm{k,f}} = \sum \frac{1}{2} m_i V_i^2 = \frac{\delta}{3}\left(\frac{1}{2} m V_{\mathrm{t}}^2\right)$$

and the internal (interaction between particles) potential energy

$$E_{\mathrm{p,\,interaction}} = \sum_{i=1}^{n} \sum_{j \neq i}^{n} E_{\mathrm{p},\,(i-j)}$$

The internal energy U is the sum of the internal kinetic energy of fluctuation and the potential interaction energy.

$$U = E_{\mathrm{k,f}} + E_{\mathrm{p,\,interaction}} = \frac{\delta}{3}\left(\frac{1}{2} m V_{\mathrm{t}}^2\right) + \sum_{i=1}^{n} \sum_{j \neq i}^{n} E_{\mathrm{p},\,(i-j)} \tag{8}$$

The apparent mechanical energy E_{app} of a system is the sum of the apparent kinetic energy $E_{\mathrm{k},G} = \frac{1}{2} m v_G^2$ and the apparent potential energy $E_{\mathrm{p},G}$:

$$E_{\mathrm{app}} = E_{\mathrm{k},G} + E_{\mathrm{p},G} = \frac{1}{2} m v_G^2 + E_{\mathrm{p},G} \tag{9}$$

The total energy E of the system is the sum of the apparent mechanical energy E (apparent kinetic energy $E_{\mathrm{k},G} = \frac{1}{2} m v_G^2$ and the apparent potential energy E_{p}) and the internal energy U. The total energy of the system E is as follows.

$$E = E_{\mathrm{app}} + U \tag{10}$$

The external medium can supply energy to the system thus modifying its total energy E. The amount supplied can be positive or negative, with organized or fluctuation work. The former comes from the work done by macroscopically moving forces; it is called organized energy work or W.

The second is usually the work done by forces acquiring random movements, i.e., with low and aleatory displacements; it also derives from the fluctuation energy produced by the shocks between external and internal particles and be called fluctuation energy or heat Q.

Energy conservation law:

$$\Delta E = \Delta\left(E_{\mathrm{app}} + U\right) = Q + W \tag{11}$$

In other words, this expression reflects the first law of thermodynamics (law of conservation of energy).

An isolated thermodynamic system is described as a system where there is no fluctuation energy or heat exchange (Q) with the external environment. The changes in the system are then adiabatic.

Thermodynamic Balance and Quasi-Static Transformations

A system reaches its thermodynamic equilibrium once it attains the different thermal, mechanical, and chemical balances. Thermodynamic equilibrium is determined through thermodynamic potential as the Helmholtz free energy at a constant temperature and volume or the Gibbs free energy for the system at a constant pressure and temperature.

The transformation of a system can sometimes occur too quickly and the particles in a system reach the level of a completely random, isotropic, homogeneous fluctuation velocity. This is then called "non quasi-static transformation." In contrast, a system follows the so-called quasi-static transformation when the change over time of the system, and a possible interaction with the external environment, allows the identification of its thermodynamic parameters at any given moment of the transformation. It is then possible to define the infinitesimal changes throughout the entire transformation.

Direction of Change: Achievable and Unfeasible Transformations

A system that is completely isolated from the external medium tends towards its intrinsic thermodynamic equilibrium. A system interacting with the external environment generally tends towards the thermodynamic equilibrium of both the system and the external medium.

No change is possible without respecting energy conservation; this is called "the first law of thermodynamics."

Not all transformations that respect the first law of thermodynamics are possible; only certain directions of transformation are feasible.

Thus, a system that is completely isolated from the outside with an initial absence of homogeneity in terms of concentration must inevitably tend towards thermodynamic equilibrium, resulting in a homogeneous composition. Inside an isolated system, an inverse trend that increases the concentration gradient is not spontaneously possible. A similar situation is observed in terms of temperature: a system that is initially nonhomogeneous in terms of temperature and is perfectly isolated from the external environment must inevitably tend towards thermodynamic equilibrium, i.e., where there is the same temperature at any point in the system. The inverse tendency is not (spontaneously) possible.

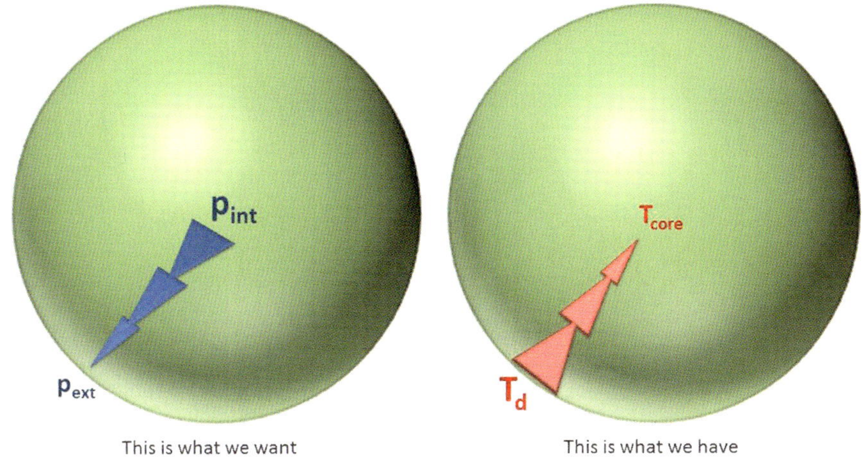

Fig. 6 Vapor transfer (*left*) and heat transfer (*right*) as they are supposed to occur during the final stage of drying

$$(p_{\mathrm{w}}/T)\vec{\nabla}_{\mathrm{W}} = -D_{\mathrm{v/s}}\vec{\nabla}\, p_{\mathrm{w}}/T \tag{30}$$

By assuming a one dimension r of a spherical particle:

$$\frac{(p_{\mathrm{W}}/T)}{\rho_{\mathrm{s}}} v_{\mathrm{v}} = -D_{\mathrm{v/s}}\left(\frac{\partial(p_{\mathrm{W}}/T)}{\partial r}\right) \tag{31}$$

$$-\lambda\frac{\partial^2 T}{\partial r^2} + \psi M L_{\mathrm{W}}\frac{\partial}{\partial t}\left[\frac{p_{\mathrm{W}}}{RT}\right] = 0 \tag{32}$$

As the exchange surface defined by $r = R$ has the highest temperature, and as the values of p_{W}/T increase with the temperature, p_{W}/T has the highest value at $r = R$:

$$\frac{\partial(p_{\mathrm{w}}/T)}{\partial r} > 0 \tag{33}$$

In this type of diffusion, vapor is schematically flowing from $r = R$ towards $r = 0$:

Figure 7

$$v_{\mathrm{v}} < 0 \tag{34}$$

However, as long as the porous matrix is full of water, the values of vapor pressure p_{W} are higher at the exchange surface than in the granule core and the surrounding medium. This results in a double motion of these vapors. Part of the

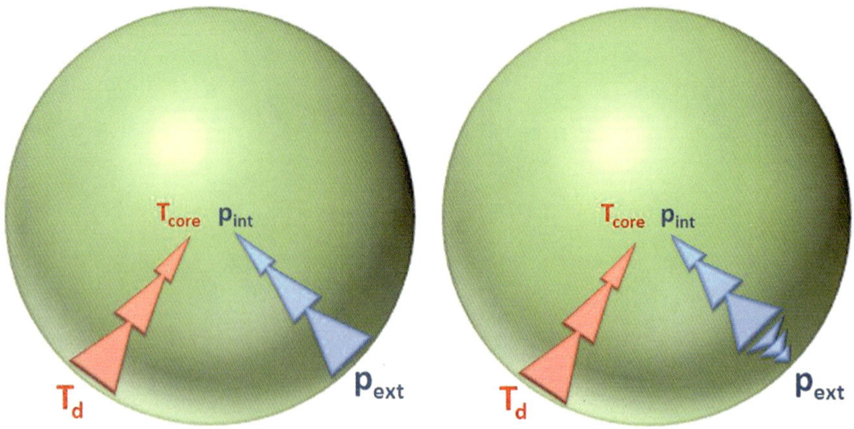

Fig. 7 The paradoxical situation of the vapor phase transfer during drying: schematic presentation (*left*) and impact in terms of progressive front kinetics (*right*)

vapor has to be transferred towards the surrounding medium. Thus, in the final period of standard airflow drying, drying is achieved through "front progression" kinetics.

There is also a specific situation where part of the vapor is transferred from the external surface towards the core of the material. This paradoxical state is completely opposite to what is required for drying (Al Haddad 2007; Al Haddad et al. 2008).

A similar paradoxical situation is obtained with freeze-drying at both sublimation and (more effectively) desorption periods.

5 Intensification of Airflow Drying: Applications

To carry out an *appropriate drying operation*, one has to intensify the entire airflow drying process. Thus, adequate external conditions for airflow can be studied, defined, and optimized as a first intensification. The natural structure has a very low effective diffusivity and this is even truer after shrinkage; a few changes to the product itself have to be made to intensify the second stage of drying. The third drying stage, which is the paradoxical stage, would require a total modification of drying technology to overcome this problem.

5.1 *First Intensification of External Transfers*

Heat and water can be transferred between the matrix and its environment in opposite directions. Heat can be transferred by contact, convection, and/or radiation, or provided by microwaves. In the airflow drying process, the heat flux depends on the temperature and velocity of the surrounding medium (air). Thus, the higher the air temperature, the higher the gradient of vapor pressure in the surrounding medium. However, this temperature increase must be limited to avoid thermal degradation or an increase in superficial crusting.

Another important airflow parameter is the humidity content, which must be kept as low as possible, although this depends on biological material and technical resources. By using appropriate desiccators or by lowering the temperature before heating (e.g., with a heating pump system), this important airflow parameter can be also controlled. However, an airflow that is too dry may induce a specific degradation of the product surface both in terms of quality and functionality aspects.

Nevertheless, airflow velocity is the main parameter that has to be properly controlled to safely obtain an appropriate external intensification. The higher the rate v, the higher the Reynolds's number Re, the higher the heat and mass convection coefficients h and k, respectively. Conversely, external intensification is no longer relevant when the internal Fick's diffusion mass transfer becomes the limiting process. Thus, the impact of airflow rate stops once it reaches a critical level v_l corresponding to the critical value k_l of the mass convection coefficient:

To prevent the external convection transfer becoming the limiting process:

$$\rho_w\, v_W\, A\big]_{r=r_{\text{surface}}} < kA\left[p_W(T_s) - p_{W,\text{air}}\right] \tag{35}$$

$$-D_{\text{eff}}\, A\,\frac{d\rho_w}{dr}\bigg]_{r=r_{\text{surface}}} < kA\left[p_W(T_s) - p_{W,\text{air}}\right] \tag{36}$$

$$k > k_l = \frac{-D_{\text{eff}}\,\frac{d\rho_w}{dr}\Big]_{r=r_{\text{surface}}}}{p_W(T_s) - p_{W,\text{air}}} \tag{37}$$

5.2 *Second Intensification*

5.2.1 Thermomechanical Texturing

After having defined the external intensifying conditions, chosen the air temperature to preserve the quality, adjusted the air humidity for convenience reason, and established airflow velocity such that k is limited by k_l, the question is: *how can we intensify internal heat and mass transfer.*

Numerous studies have demonstrated a much higher thermal diffusivity than effective water diffusivity; this phenomenon is more effective with more compact materials after shrinkage. Only freeze-drying with highly porous materials may involve the opposite situation, where the internal transfer of heat by conduction is the limiting process.

Knowing that plants and other porous materials usually have a much higher heat diffusivity than Fick's diffusion diffusivity ($\alpha \gg D_{eff}$), the question becomes: how can we increase both *starting accessibility* and *effective diffusivity*? Putting aside grinding, texturing operations should be included as a further (second) intensification. These should improve both the airflow drying *performance* and the *quality* of the end product.

The expansion of a porous structure enhances water diffusivity, regardless of the liquid or vapor phase. Thus, expansion by DIC is particularly appropriate at this stage of drying.

Instant controlled pressure drop (DIC) technology is distinguished by its capacity to handle an extremely wide range of food products, regardless of their sensitivity to heat. With this kind of controlled texture expansion, the final stage of drying can be reached more quickly, thus reducing energy consumption and consequently the manufacturing cost.

"Swell-drying" is a special drying process coupling airflow drying to DIC expansion. DIC is often used just after an initial conventional drying step using hot air to address the compactness of the product through a controlled expansion. This operation is generally achieved when the moisture content results in the glass transition of the material at ambient temperature. The drying kinetics will thus be intensified with a consequent reduction in energy consumption and a marked improvement in the quality of the final product.

5.2.2 Swell-Drying or How to Include DIC Treatment in an Airflow Drying Process?

Conventional airflow drying, sun drying, solar drying, vacuum drying, etc., are known to be easy to use, they can be tightly controlled, and are widely used. Sun and solar drying can be employed to directly process fruit and vegetables near the harvesting area, which is suitable for many products. In addition there is the advantage of using a free and renewable energy, often significantly reducing the energy cost of the operation. However, low organoleptic and nutritional quality, hygiene problems, and difficulties in downstream processing of the finished product (e.g., grinding, rehydration) constitute significant barriers to a more widespread use.

The most modern drying methods, such as superheated steam and vacuum microwave drying (MWD), have improved the kinetics without, however, resolving the issue of quality in terms of microbiological decontamination, texture control, and powder production. In addition, the high cost of freeze-drying and the loss of some quality vectors, such as flavor and color, limit its use to products with high added value such as pharmaceuticals. Intermittent drying processes have been used as alternative drying methods, in the main for biomaterials. Various parameters can be selected and modified such as temperature, airflow rate, operating pressure, as well as heat input. Drying kinetics and the quality of dried bio-products have been improved (Chua et al. 2003). The effect of temperature variation on product quality has been established. Indeed with a proper choice of the air temperature for intermittent drying, a larger quantity of certain vitamins can be obtained than under isothermal conditions. The low quality of hot air- or solar-dried products is mainly due to the texture, which is too compact as a result of shrinkage. Indeed, when glass transition does not take place, the product loses water and collapses under its own weight. This compact structure dramatically reduces mass transfer diffusivity, thus causing a significant thermal degradation with loss of vitamins, color, and, of course, flavor.

Swell-drying involves coupling airflow drying with DIC texturing. This process has been used mainly but not exclusively with plants, fruit, vegetables, seaweeds and microalgae, seafood, and meat products. The reduction in processing time leads to a significant improvement in product quality. The vitamin content of the finished product is higher than that of conventionally dried products, although not as high as in freeze-dried products. However, the availability of flavonoids and the antioxidant activity are often markedly higher than in the raw material itself. The DIC-expanded texture also makes it possible to preserve color and flavor. In addition, the parameters of DIC treatment can be defined to achieve microbiological decontamination, which is extremely important. Thus, DIC products normally have a much longer lifetime because in the absence of insects, larvae and other factors, the storage time, can easily exceed 1 or even 2 years.

5.2.3 Specificity of DIC Treatment

DIC treatment can be used to expand different materials by an appropriate autovaporization that depends on the amount of vapor generated, the internal pressure constraints, and the pressure drop rate, as well as the viscoelastic behavior of the material. The latter is a function of water content and temperature. However, another important question is the following: how can we get adequate viscoelastic behavior while preserving the expanded structure once the pressure drops instantaneously?

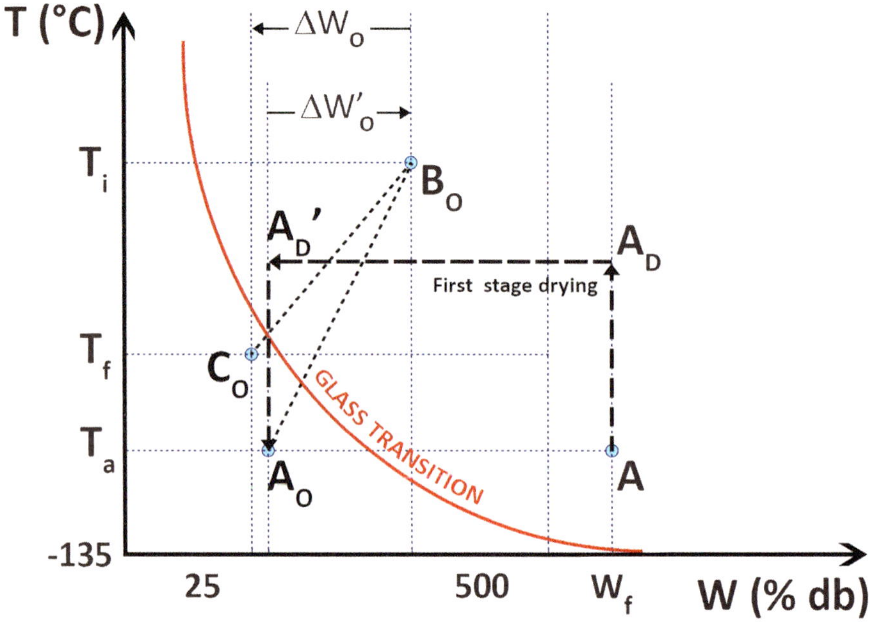

Fig. 8 Temperature/moisture content dry basis during the first stage of drying ($AA_D\,A'_D\,A_o$) and DIC treatment ($A_oB_oC_o$)

DIC is usually used to dry many porous hygroscopic materials such as minerals, food, and standard biological materials, when the water content is so low that the material is close to glass transition at equilibrium temperature. Except for some categories of products (meat, fish), a first stage of drying is usually needed before the DIC expansion (Fig. 8).

The impact of expansion by DIC can result in four response parameters:

1. The quantity of moisture at final equilibrium W_e expressed in g H_2O/100 g dry basis db often decreases with expansion, as does water activity.
2. Water diffusivity D_{eff} expressed in $m^2\ s^{-1}$ is calculated in the range excluding the initial (surface interaction) and the paradoxical phase. It increases significantly with porosity and expansion ratio. After an optimized DIC treatment, D_{eff} was often 2–10 times higher.
3. Starting accessibility δW_s (expressed in g H_2O/100 g db) significantly increases with the effective surface exchange, which is closely related to the specific surface area and expansion ratio; this is twice as high, and sometimes even higher, than with the raw material.
4. Drying time, due to an increased effective diffusivity and starting accessibility, is often reduced by more than 50 %. In numerous cases, drying time with DIC is 1–2 h compared with 6–12 h for samples treated using conventional drying processes.

5.3 *Third Intensification: Overcoming the Paradoxical Phase*

The final stage of airflow drying is usually a paradoxical stage where gradients of both temperature (as the driving force of heat transfer that occurs through a conduction-type diffusion) and vapor pressure (as the driving force of vapor transfer that occurs through a Fick's-type diffusion) are oriented from the surface towards the core of the product. Indeed, in the final stage of airflow drying, the residual water, either free or bound, is merely transferred within the porous matrix in a vapor phase. This is more paradoxical in the case of drier, more porous materials in that it induces low conductivity and therefore a higher temperature gradient.

However, it is possible to overcome this problem and notably improve the kinetics by adopting more appropriate heating methods than conduction and more effective mass transfer than that described by Fick's law. Thus, mass transfer according to Darcy's law with the total pressure gradient as the driving force is a real solution.

Hence, the best operations for a drying stage after DIC are those that are fundamentally able to overcome the paradoxical stage of airflow drying. They are those which ensure a deep uniform heating (such as microwaves—MW). They are also those that remove the residual water, either free or bound, from the porous matrix through Darcy's law of vapor transfer; here the driving force is the total pressure gradient (such as multi-flash drying—MFD or super heated steam drying—SHSD). MW heating does not involve any notable internal temperature gradient between the surface and the core of the material. The main cause of the paradoxical situation is reduced or even removed. Although MW is very prejudicial for the first two stages of drying because of the degradation in the quality and the high cost of the equipment, MW can be used effectively at the final drying stage.

5.3.1 Multi-flash Drying

Heating with Compressed Air

The second proposal is MFD, carried out using a specific DIC autovaporization with compressed air instead of saturated steam; it involves a Darcy-type transfer

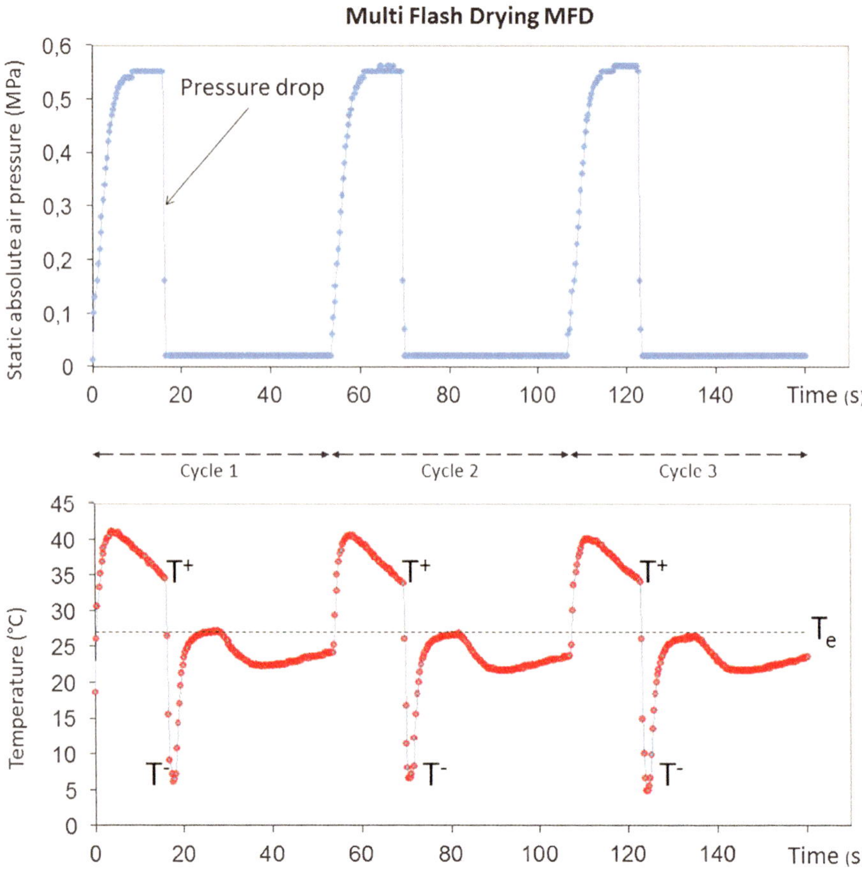

Fig. 9 Pressure and temperature evolution vs. time during MFD treatment

of vapor from inside the material towards the surrounding medium that is under a vacuum.

MFD contains *n* repetitions of heating cycles and pressure drops (Fig. 9).

The first DIC heating stage, starting from an initial product temperature T_i, establishes a uniform treatment temperature T_t that is generally much lower than the boiling temperature of water at the treatment pressure. Heating with pressurized air (or neutral gas) quickly increases the temperature of the product surface (exchange surface) to the temperature of the air, mainly by convection. This is followed by a conduction process within the product, with an effective conductivity λ_{eff} that is much higher than the calculated or measured static value λ because of the internal evaporating/condensation phenomenon.

The quantity of heat absorbed by the product is estimated as:

Sensible heat Q at constant volume:

$$Q = m_{\mathrm{d}}\left(c_{\mathrm{v,d}} + Wc_{\mathrm{v,w}}\right)\left(T_{\mathrm{t}} - T_{\mathrm{i}}\right) \tag{38}$$

where:

$$c_{\mathrm{v}} = \sum_{i=1}^{n} \left(C_i c_{\mathrm{v},i}\right) \tag{39}$$

$$c_{\mathrm{v}} = \frac{1}{m}\frac{\partial Q}{\partial T}\bigg]_{\mathrm{v}} \tag{40}$$

m_{d} is the dry weight of the sample; c_{v} the specific heat at a constant volume of dry material, m_{d}, and water, W. It is expressed in J kg^{-1} K^{-1}

Since the superficial temperature is quickly established at T_{t}, heat transfer within the product begins immediately. Heat transfer occurs within the product according to a Fourier-type law:

$$\vec{\varphi} = -\lambda_{\mathrm{eff}} \cdot \vec{\nabla}\, T \tag{41}$$

where λ_{eff} is the effective conductivity including evaporation/condensation phenomenon within the porous material (W m^{-1} K^{-1})

By adding the continuity ratio:

$$\frac{\partial T}{\partial t} = -\vec{\nabla} \cdot \left(\alpha_{\mathrm{eff}} \vec{\nabla}\, \rho_{\mathrm{W}}\right) \tag{42}$$

Both effective diffusivities α_{eff} and D_{eff} expressed in m^2 s^{-1} depend on water content W (dry basis). We usually use the Dickerson approximation (Dickerson 1969) for estimation of effective thermal diffusivity:

Estimation of effective thermal diffusivity α (m^2 s^{-1}):

$$\alpha_{\mathrm{eff}} = \frac{8.8\ 10^{-8} + W\alpha_{\mathrm{W}}}{1 + W} \tag{43}$$

Effective thermal diffusivity α_{eff} is also slightly dependent on the temperature T, the porosity ratio, and the average size of pores. Although thermal diffusivity α_{eff} depends on the material and its structure, temperature, and water content, the values remain at about 10^{-7} m^2 s^{-1}.

Pressure Drop Stage

By abruptly dropping the pressure, an adiabatic autovaporization of the overheated water occurs, inducing an instant cooling of the residual material.

The structure can be partially preserved if the cooling stage and a lower water content allow the product to cross the glass transition border. The amount of water removed from the product by autovaporization through the exchange surface becomes:

m_v:

$$m_v = m_d \frac{\left(c_{p,d} + W c_{p,w}\right)\left(T_p - T_t\right)}{L} \tag{44}$$

where T_p is the temperature of the product just after the pressure drops.

Depending on the pressure drop time, T_p can be the equilibrium level T_e (quasi-static transformation):

For a long pressure drop time (quasi-static transformation):

$$T_p = T_e \tag{45}$$

T_p can also be much lower than the equilibrium temperature T_e when the pressure is dropped instantaneously (instantaneous transformation):

For an instant pressure drop (instantaneous transformation):

$$T_m \leq T_p \leq T_e \tag{46}$$

Thus, just after dropping the pressure, the water content in the product decreases (%, dry basis):

Modification of water content (% dry basis)

$$\Delta W = \frac{\left(c_{p,d} + W c_{p,w}\right)\left(T_p - T_t\right)}{L} < 0 \tag{47}$$

After the instant pressure drop step and during the vacuum step, the total internal pressure is mainly due to vapor in the porous medium. Under DIC conditions, vapor pressure in the porous material just after the drop in pressure is assumed to be much higher than the external pressure (Allaf 2009), and mass transfer is achieved by permeability, according to a Darcy-type process, with the total pressure gradient as the "driving force":

General Darcy law:

$$\rho_v\left(\vec{v}_v - \vec{v}_d\right) = -\frac{K}{\nu_v}\vec{\nabla} P \tag{48}$$

The permeability K expressed in m^2 mainly depends on the structure, porosity, and tortuosity of a solid matrix. The kinematic viscosity ν_v expressed in $m^2\ s^{-1}$

depends on the nature of the transferred fluid (density, molecular size). Sulaiman (2013) measured the expansion time to be between 20 and 200 ms, depending on the material and the autovaporization conditions. By assuming that $\vec{v_d} = 0$, and the transfer to be a one-dimensional radial within a spherical material, (48) becomes:

$$\rho_v \vec{v}_v = -\frac{K}{\nu_m}\frac{\partial P}{\partial r} \tag{49}$$

Therefore, the transfer of volatile molecule vapor such as water and essential oils within an expanded granule is assumed to be from the core towards the surrounding medium through the gradient of total pressures. By using the first Darcy-type law describing this transfer, coupled to mass conservation and continuity, and by integrating between the hole (whose radius is R_o) and the external radius (R_s) of the spherical shape, (49) becomes:

$$\dot{m}_v = \frac{4\pi K}{\nu_m}\frac{\left(P_{\text{globule}} - P_{\text{ext}}\right)}{\left(\frac{1}{R_o} - \frac{1}{R_s}\right)} \tag{50}$$

The total pressure (P_{globule}) in the hole decreases with time (t), depending on the flow ($\dot{m}_m$). The globule radius (R_o) is assumed to be constant (absence of expansion) as is the temperature, which would appear to remain constant at the value reached just after the pressure drop, defined by the level of the vapor temperature–pressure equilibrium. Equation 50 becomes:

$$\frac{dP_{\text{globule}}}{dt}\frac{MV_{\text{hole}}}{RT} = -\frac{4\pi K}{\nu_v}\frac{\left(P_{\text{globule}} - P_{\text{ext}}\right)}{\left(\frac{1}{R_o} - \frac{1}{R_s}\right)} \tag{51}$$

$$\left(P_{\text{globule}} - P_{\text{ext}}\right) = \left(P_{\text{globule},o} - P_{\text{ext}}\right)\exp\left(-\frac{3KRT}{\nu_v}\frac{t}{MR_o^3\left(\frac{1}{R_o} - \frac{1}{R_s}\right)}\right) \tag{52}$$

Change vs. time of vapor amount:

$$\dot{m}_v = \dot{m}_{v,o}\exp\left(-\frac{3KRT}{\nu_m}\frac{t}{MR_o^3\left(\frac{1}{R_o} - \frac{1}{R_s}\right)}\right) \tag{53}$$

Characteristic time of Darcy's mass transfer:

$$\tau = \frac{MR_o^3\left(\frac{1}{R_o} - \frac{1}{R_s}\right)\nu_m}{3KRT} \tag{54}$$

Usually, the time t_v of the vacuum stage (f) is sufficient to allow the vapor to be transported towards the surrounding medium of the vacuum tank.

Thus, DIC provides a very effective solution to the problems posed by the paradoxical phenomenon. It uses autovaporization instead of evaporation, and total pressure gradient instead of the normal diffusion phenomenon. MFD towards a vacuum is a highly appropriate way of overcoming this paradoxical situation. Between the initial internal vapor pressure of 25 kPa at a temperature of about 65 °C and an external total pressure that is usually at about 3–5 kPa, a total pressure gradient is established, thus leading to a highly effective Darcy-type vapor transfer.

5.3.2 Other Methods to Address the Paradox

Another method to address this paradox and to intensify the drying process is to heat DIC-expanded samples with microwaves. This deep heating allows a relatively effective Darcy-type transfer between the internal and external media. As the moisture to be removed is usually less than 20 kg/100 kg of dry material, the microwave power needed, and thus the cost of the equipment, is significantly lower than when samples are dried using only microwaves (more than 800 kg/100 kg of dry material).

Finally, superheated steam can also solve the paradoxical problem. Superheated steam is steam that has a temperature above the saturation or boiling point. At this point, the steam can transfer heat to the product, which almost reaches the temperature of steam. Internal vapor pressure is then higher than the total external pressure. Darcy's transfer of vapor is then carried out much more effectively and drying the remaining water is achieved in a much shorter time.

These three processes involve a possible total internal pressure that is higher than the external pressure, leading to a transfer similar to that of Darcy's law: the higher the total pressure gradient, the faster the vapor transfer and the drying kinetics. This allows an effective saving in energy consumption as well as a marked increase in the quality of the end product.

6 Microstructure and Quality of Swell-Dried Fruits and Vegetables

There are many quality attributes which are related to the physical state of the dried ingredients. Minor components such as flavors, vitamins, and enzymes in the dried product are encapsulated into the major amorphous ingredient matrix. Any change in the physical state of the product can affect the physicochemical characteristics of all the ingredients of the product.

Drying causes irreversible structural damage to the cellular structure of foods. The structure of dried foods depends on the drying methods and conditions such as temperature, relative humidity, airflow velocity, and the initial physicochemical properties of the foods. Information on the structural properties of dried food products is needed to design the process, estimate other properties such as thermal conductivity, density, and moisture diffusivity, and determine food quality (Rahman 2001; Rahman and Sablani 2003). During any drying operation one critical entity, which competes with process economics, is quality preservation since most dehydrated foods are produced by airflow drying.

Observations by scanning electronic microscopy (SEM) of various food products like cheese snacks and powder, caseinate powder, beef meat, chicken breast, seafood, fish, and fruit powders illustrate the direct impact of DIC on the structure of dried products. In general, airflow-dried products tend to be of poor quality due to excessive thermal damage and additionally they have a reduced reconstitution capacity due to the structural disruption incurred during drying, e.g., shrinkage and internal porosity (Sandhu and Parhawk 2002; Pimpaporn et al. 2007; Prachayawarakorn et al. 2008). The structural changes that occur after DIC texturing explain the higher quality: improved rehydration capacity, increased WHC, lower a_w, and improved sensory and nutritive values. Kerdpiboon et al. (2007) reported significant physical changes in terms of percentage of shrinkage and rehydration ratio of carrot and potato samples after airflow drying. These apparent changes correlated well with their microstructural changes.

Swell-dried products have an open cell structure with a connective porous network that is completely different to that of airflow-dried products, which have a compact, dense structure with a hard outer crust.

6.1 Flavor, Aroma, Taste, Color Preservation, and Crispy Texture

Airflow-dried products can frequently suffer from color and flavor changes that usually occur during the long final drying period (Karathanos 1993; Bhandari and Howes 1999). Some authors attributed poor aroma retention to the behavior of volatiles in the rubber matrix (Roos and Karel 1991; Whorton and Reineccius 1995). However, experiments proved that the final drying period, because it is a time-consuming stage, involves a considerable loss of aroma. Since the final stage of swell-drying is significantly shorter thanks to a higher effective diffusivity, swell-dried fruits and vegetables are distinguished by a marked improvement of sensory properties such as color, flavor, aroma, and crispness. The relative expansion ratio of swell-dried vegetables such as onion, carrot, and potatoes is 200–300 % compared to control samples. Swelling or expansion of the product results in an increase in the specific surface area, which is twice that of swell-dried apples compared to airflow-dried apples (Mounir et al. 2011a). Iguedjtal et al. (2008) also reported this behavior for swell-dried potatoes.

Crunchy snacks are popular products in many countries. The texture of dried products is related to glass transition T_g and systematically depends on structure and

Fig. 10 Swell-dried fruits and vegetable

shrinkage (porosity) (Mounir et al. 2011b). The brittleness of a product with low moisture content can often be described as crispy and crunchy, which are sensations related to the fracture properties of food products; they are the most important characteristics that define customer preferences in snacks (Farroni et al. 2008).

Fig. 11 DIC textured French cheese

Empirical models of crunchy and crispy texture measurements are determined as a function of DIC operating parameters (Alonzo-Macías 2013). These different criteria are used to optimize DIC processing parameters, thus producing high-quality textured dried fruits and vegetables such as potatoes, carrots, onions, green beans, tomatoes, strawberry, and pepper (Fig. 10).

Even for very sensitive products, such as pure cheese, multi-criteria optimization of DIC processing parameters resulted in the production of high-quality cheese snacks and powder in a tightly controlled process. Pure French cheese can be significantly expanded to produce pure cheese snacks without any additives (Mounir et al. 2011c). The expansion ratio of swell-dried cheese snacks was as high as 3,500 % (Fig. 11). Similar results were obtained with chicken breast meat and sodium caseinate; the expansion ratio was 15 and 18 times higher than untreated samples, respectively. After grinding, these products became "expanded granule powders" which are a new type of powder with very high functional properties.

Another attractive application is the processing of fresh meat (beef, chicken), fish, and seafood. It is specific because DIC texturing is applied directly on fresh material followed by a short mild airflow drying.

6.2 Nutritive Value

6.2.1 Decontamination

DIC treatment can be used as a decontamination process with a large variety of food, animal feed, powders, baby foods, intermediate ingredients, etc. It can also serve as a superficial decontamination for fresh cut fruit and vegetables. DIC often combines swell-drying with a perfect decontamination.

Decontamination by DIC is not limited to the vegetative form of microorganisms since spores can also be destroyed. Albitar et al. (2011) showed that DIC efficiently decreased the initial floral load of onions. The *log* of microorganism load was found to be close to 1 compared with 4 for swell-dried and airflow-dried onions, respectively. Effective decontamination was obtained with under 0.4 MPa DIC steam pressure for 15 s. The results showed that both steam pressure and thermal treatment time had a significant effect in decreasing the initial load: the higher the steam pressure and the longer the treatment, the higher the *log* decontamination index.

To better understand how DIC works as a decontamination process, studies were carried out to evaluate the impact of multiple cycles of pressure drops towards a vacuum on vegetative forms and spores. We observed that, for conventional thermal treatments, including ultrahigh temperature (UHT), temperature and treatment time were the only operating parameters. For DIC the number of pressure drop cycles is an additional effective parameter, acting on both vegetative flora and spores.

The high destruction rate of vegetative flora and spores with DIC is probably due to the various stresses resulting in irreversible changes in the cells of microorganisms such as:

1. Protein denaturation
2. Shrinkage of the cellular membrane
3. Denaturation of protein enzymes
4. Breakdown of the cell structure and/or
5. Instant thermal and mechanical stress

These changes are mainly due to the heat shock to which microorganisms are exposed as the temperature increases; indeed, this shock increases after several cycles of pressure drops that induce successive cycles of heating-instant cooling. Moreover, they may be due to the stress and strains exerted on the cell after steam expansion inside the cell followed by an instant pressure drop towards a vacuum, resulting in the breakdown of cell structure.

6.2.2 Preservation and Increasing Availability of Bioactive Molecules

Following conventional airflow drying, fruit and vegetables, such as apple, onion, and super fruits, lose most of their nutritive value, especially in terms of bioactive compounds such as antioxidant molecules. This mainly occurs during the time-consuming final period, principally because of shrinkage, which results in very low diffusivity. In contrast, since DIC texturing significantly increases diffusivity and reduces drying time, it preserves the natural nutritive value and, by opening the structure and possibly rupturing cell walls, markedly improves the availability of antioxidant compounds (Mounir et al. 2011a). Thus, DIC/swell-drying can not only preserve these compounds but also increase their availability. As an example, the amount of quercetin in DIC-treated apple compared to untreated fresh samples increased by up to 700 %.

Similar results were obtained with other fruits (e.g., cranberry fruits and seeds, sumac) and vegetables. Furthermore, the results with onions proved that the instantaneity of the pressure drop towards a vacuum is an important factor that influences the availability of flavonoids.

Finally, DIC/swell-dried snacks are acrylamide-free snacks, in contrast to conventional fried snacks.

6.2.3 Improving Digestibility, Elimination of Anti-nutritional and Allergic Factors

Legumes such as seeds and beans (soya, lupine, rapeseed, etc.) are the main sources of plant protein. However, the presence of anti-nutritional factors limits their use in food because of digestive and allergenic problems. DIC heat treatment can reduce and even destroy the anti-nutritional factors in soybeans and lupines through its thermal impact on phytates, with a higher efficacy and more quickly than conventional heat treatments (De Boland et al. 1975; Haddad and Allaf 2007; Haddad et al. 2007).

Similar results were obtained with soybean trypsin. From these results it was possible to scale up DIC to the industrial level to decrease the anti-nutritional factors (Haddad and Allaf 2007). This has the advantage of using a moderate temperature for a shorter time (6 min) compared with many other processes cited in the literature (Baker and Mustakas 1973; Rackis et al. 1986).

6.3 Functional Properties

6.3.1 Product–Water Interactions

Desorption and Adsorption Isotherms

Although the drying kinetics was better and the starting accessibility and effective diffusivity higher, water was better bound in DIC-expanded material. Indeed, as the specific surface area increases, the number of polar groups on the external surface also increases and they thus react with more water molecules. The higher quantity of bound water leads to lower water activity (a_w), reduces water availability for microorganisms, and inhibits certain chemical reactions (Arslan and Togrul 2006).

Comparing DIC/swell-dried products with airflow-dried products, Iguedjtal et al. (2008) and Albitar et al. (2011) found that swell-dried potatoes needed a higher water content to reach the same water activity. These higher values for adsorption in swell-dried potatoes were in line with the porous structure, which increased by 45 % after DIC texturing. Similar results were obtained by Iguedjtal et al. (2008), Abdulla et al. (2009), and Téllez-Pérez (2013). Furthermore, the results obtained were corroborated by those recorded by Mazza (1982) who compared the adsorption isotherm of freeze-dried and vacuum-dried potatoes.

a_w is an extremely important parameter that can describe the thermodynamic equilibrium state of water and predict quality, stability, shelf life, and storage conditions. Moreover, it is essential in the design and optimization of drying equipment (Al-Muhtaseb et al. 2004; Durakova and Menkov 2004; Katz and Labuza 1981; Kaymak-Ertekin and Gedik 2004). Thus, reducing water activity in a dry or dehydrated product normally maintains the proper product structure, texture, stability, density, and rehydration properties during storage. So from the results obtained, it can be concluded that swell-dried products are more stable than CHA-dried products with the same water content.

Rehydration Kinetics and Water Holding Capacity

In different cases, the kinetics and rehydration capacity of well-optimized swell-dried materials were much greater than for airflow-dried products, although usually slightly lower than for freeze-dried products. For instance, the rehydration behavior of chicken breast meat significantly depends on drying methods (airflow drying, freeze-drying, and swell-drying) (Mounir and Allaf 2012). The poor rehydration capacity of airflow-dried samples correlates with shrinkage and thermal degradation due to the lengthy drying process. In the case of chicken breast meat, this

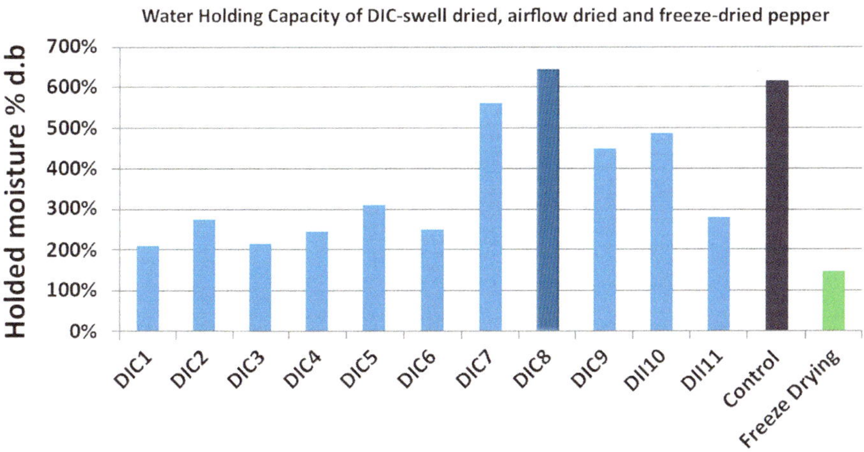

Fig. 12 Water holding capacity (WHC) of swell-dried, low temperature airflow-dried and freeze-dried samples

lengthy airflow drying denatured the myofibrillar and collagenous connective tissue proteins, resulting in a weak rehydration capacity.

Other studies were also conducted to investigate the material's behavior during rehydration. In the case of Moroccan green pepper (Téllez-Pérez et al. 2012) and strawberry (Maritza et al. 2012), freeze-dried samples had the best rehydration capacity (starting accessibility, effective diffusivity, and WHC) compared to other samples (airflow dried and swell dried). Despite such good rehydration behavior, the freeze-dried samples could not bind water strongly, revealing a weak WHC. In contrast, well-optimized swell-dried and low temperature air-dried samples had a very high WHC (Fig. 12).

Comparing the different DIC-treated materials, various research has proved that saturated steam pressure is the DIC operating parameter which has the most significant effect on rehydration capacity, and the higher the saturated steam pressure, the higher the starting accessibility and the effective diffusivity of the rehydration kinetics. Albitar et al. (2011) correlated this type of behavior in swell-dried onion and blackcurrant to their large specific surface area, which resulted in a greater WHC and a higher amount of water absorbed in a shorter time. It is normally admitted that the degree of rehydration closely depends on the sample processing conditions, preparation, composition, and the extent of structural and chemical disruption induced by drying (Singh et al. 2006). Moreover, in the work of Mayor and Sereno (2004), further causative factors for limited water uptake were suggested, namely, starch crystallinity, protein denaturation, and hydrogen bonding of the macromolecules.

This faster and more complete rehydration process also correlated with decreased drying time, absence of shrinkage, and, therefore, the presence of well-defined intercellular voids to promote an increased rehydration rate, as was assumed by Cano-Chauca et al. (2005) and Haas et al. (1974). The open cell texture obtained by DIC texturing leads to a much higher water diffusivity, decreasing both drying and rehydration times.

Thus, the impact of DIC processing parameters was statistically calculated. It was observed that saturated steam pressure generally markedly increased WHC, and to a much greater extent than thermal processing time.

This increase in the WHC of meat is due to modifications of the structure of the muscle cells and their components, as well as the amount of extracellular space within the muscle itself. The results concerning expansion ratio together with the analyses of microstructure by scanning electron microscopy explain this modification. On the other hand, in this case, the thermal treatment must have a direct impact on the change in net protein charge, allowing a greater amount of water to be retained.

6.3.2 Oil Holding Capacity

WHC and oil holding capacity (OHC) are physical properties that indicate the aptitude of a material to hold water and oil, respectively. It is very important to understand these physical properties for flour or powdered food materials as numerous food powders are usually blended together to produce a final food product.

Setyopratomo et al. (2011) found that DIC texturing of cassava flour increased OHC compared to untreated flour. OHC increased under defined DIC conditions to 2.0 g oil/g dry cassava compared with only 0.4 g oil/g dry cassava for DIC textured and CHA control cassava, respectively. The increase in OHC (up to 500 %) was related to chemical modifications combined with microstructural changes as a result of DIC thermomechanical effects.

7 Energy Consumption

Thermal drying operations are found in almost all industrial sectors and consume 10–25 % of the national industrial energy in the developed world. As the cost of energy has been increasing in recent years, the operating cost of energy-intensive operations such as thermal drying is also rising; moreover, greenhouse gas emissions are correlated with energy consumption. With the world's emerging economies rapidly industrializing, the energy consumed for thermal drying and the resulting adverse environmental impact of greenhouse gas emissions will inevitably increase over time. This puts pressure on the energy-intensive industry to look for energy-efficient technologies and to find ways to reduce energy consumption. Since drying is an energy-intensive operation, the energy consumption of an industrial drier is huge and this in turn makes the unit operation costly.

The most effective solution to this growing problem is to develop and utilize energy-efficient drying technologies that will reduce net energy consumption and diminish the impact on the environment. Conventional industrial driers usually operate at 30–70 % efficiency levels. Hence, industrial drying is definitely an area that is desperate for energy-efficient technology.

Generally, conventional driers are operated at low energy efficiency due to poor design and a lack of understanding of the fundamental scientific principles of drying. Moreover, these driers were designed when energy was plentiful and cheap. Nowadays, the scenario has radically changed and energy cost is a major concern not only from the point of view of the economy but also in terms of the environmental impact.

Swell-drying involves texturing of the material to be dried. This improves effective diffusivity and starting accessibility and decreases the time needed to reach an appropriate final water content. As the drying rate increases, in terms of rapid water removal and short drying time, the operating cost is reduced and the energy consumption diminished.

At the industrial scale, DIC treatment could require 0.2 kWh total energy consumption per kg of dried material. The total energy needed per kg of end product was determined and calculated to be 8, 7.6, 13.7, and 6.8 kWh for conventional airflow drying, MWD, freeze-drying (FD), and swell-drying, respectively (Mounir et al. 2012).

8 Conclusion

Shrinkage resulted from drying of biological materials has prejudicial negative impacts on both the quality of the material and the performance of the process. It can be addressed by using the expansion by "instant controlled pressure drop" (DIC). This results in higher exchange surface and greater diffusion of moisture through the porous structure, for both the residual dehydration process and the rehydration kinetics and capacity. The DIC-dried products are highly textural quality with great preservation of active nutritional molecules. They can be used as snacking and are considerably suitable for grinding. The resulting expanded granule powders consist of a high specific surface area, with uniform distribution of granulometry, and an appropriate structure for great functional behaviors and organoleptic quality. DIC technology can be used as a UHT treatment to decontaminate the foodstuffs. These dried, expanded, and completely decontaminated products can be used as highly nutritional "snacks," which can also be ground easily to produce high-quality expanded granule powders with excellent sensory, nutritional, functional, and textural properties.

Consumer demand and various end uses have increased demand for processed products that preserve as much of their original characteristics as possible. In industrial terms, this requires the development of unit operations that minimize the adverse effects of processing. The effect of food processing on end product quality may determine the usefulness and commercial viability of those unit process operations.

In the particular case of food drying, this indicates a loss of volatiles and flavors, changes in color and texture, and a decrease in nutritional value. Furthermore, residual enzyme activity and microbial activity in dried foods are essential parameters that affect product quality and shelf life.

The first industrial scale DIC reactor using the swell-drying process started producing dried vegetables for instant soups in 1993. In 2001, the principal activities of one start-up company, ABCAR-DIC Process in France, were DIC engineering, designing and building equipment, technology research, etc. Different DIC reactors are now operating worldwide, e.g., in France, Spain, Italy, Mexico, Malaysia, and China. They are used not only for swell-drying but also in decontamination and expansion processes, thus improving the extraction of active volatile and nonvolatile molecules, e.g., for pharmaceuticals, cosmetics, and nutrition.

References

Abdulla G, Belghit A, Allaf K (2009) Impact of instant controlled pressure drop treatment on moisture adsorption isotherm of cork granules. Drying Technol 27(2):237–247. doi:10.1080/07373930802605990

Al Haddad M (2007) Contribution théorique et modélisation des phénomènes instantanée dans les opérations d'autovaporisation et de déshydratation. Université de La Rochelle, La Rochelle

Al Haddad M, Mounir SM, Sobolik V, Allaf K (2008) Fruits and vegetables drying combining hot air, DIC technology and microwaves. Int J Food Eng 4(6):Article 9

Albitar N, Mounir S, Besombes C, Allaf K (2011) Improving the drying of onion using the instant controlled pressure drop technology. Drying Technol 29(9):993–1001. doi:10.1080/07373937.2010.507912

Allaf K (1982) Transfer phenomena and industrial applications. Lebanese University Faculty of Science, Beirut

Allaf K (2009) The new instant controlled pressure-drop DIC technology. In: Chemat F (ed) Essential oils and aromas: green extraction and application. Har Krishan Bhalla and Sons, New Delhi, pp 85–121

Al-Muhtaseb AH, McMinn WAM, Magee TRA (2004) Water sorption isotherms of starch powders: part 1: mathematical description of experimental data. J Food Eng 61(3):297–307. doi:10.1016/s0260-8774(03)00133-x

Alonzo-Macías M (2013) Comparative study of strawberry drying processes: hot air drying, freeze-drying and swell drying, application on the biological compounds preservation. University of La Rochelle, La Rochelle

Arslan N, Togrul H (2006) The fitting of various models to water sorption isotherms of tea stored in a chamber under controlled temperature and humidity. J Stored Prod Res 42(2):112–135. doi:10.1016/j.jspr.2005.01.001

Baker E, Mustakas G (1973) Heat inactivation of trypsin inhibitor, lipoxygenase and urease in soybeans: effect of acid and base additives. J Am Oil Chem Soc 50(5):137–141. doi:10.1007/bf02640466

Bhandari BR, Howes T (1999) Implication of glass transition for the drying and stability of dried foods. J Food Eng 40(1–2):71–79. doi:10.1016/s0260-8774(99)00039-4

Bonazzi C, Bimbenet JJ (2003) Séchage des produits alimentaires - Principes. Techniques de l'Ingénieur F-3:1–14

Cano-Chauca M, Stringheta PC, Ramos AM, Cal-Vidal J (2005) Effect of the carriers on the microstructure of mango powder obtained by spray drying and its functional characterization. Innov Food Sci Emerg Technol 6(4):420–428. doi:10.1016/j.ifset.2005.05.003

Changrue V (2006) Hybrid (osmotic, microwave-vacuum) drying of strawberries and carrots. McGill University, Montreal

Chen R-H, Chow LC, Navedo JE (2002) Effects of spray characteristics on critical heat flux in subcooled water spray cooling. Int J Heat Mass Transf 45(19):4033–4043. doi:10.1016/s0017-9310(02)00113-8

Chua KJ, Chou SK, Ho JC, Mujumdar AS, Hawlader MNA (2000) Cyclic air temperature drying of guava pieces: effects on moisture and ascorbic acid contents. Food Bioprod Process 78(2):72–78. doi:10.1205/096030800532761

Chua KJ, Mujumdar AS, Chou SK (2003) Intermittent drying of bioproducts—an overview. Bioresour Technol 90(3):285–295. doi:10.1016/s0960-8524(03)00133-0

Crank J (1975) The mathematics of diffusion. Oxford University Press, Oxford

De Boland AR, Garner GB, O'Dell BL (1975) Identification and properties of phytate in cereal grains and oilseed products. J Agric Food Chem 23(6):1186–1189. doi:10.1021/jf60202a038

Dickerson RW (1969) Thermal properties of foods. In: Tressler DK, Van Arsdel WB, Copley MJ (eds) The freezing preservation of food, vol 2, 4th edn. AVI Publishing Company, Westport, pp 26–51

Donsì G, Ferrari G, Nigro R, Di Matteo P (1998) Combination of mild dehydration and freeze-drying processes to obtain high quality dried vegetables and fruits. Food and Bioproducts Process 76(4):181–187. doi:10.1205/096030898532061

Durakova AG, Menkov ND (2004) Moisture sorption characteristics of rice flour. Food Nahrung 48(2):137–140. doi:10.1002/food.200300387

Farroni AE, Matiacevich SB, Guerrero S, Alzamora S, del Pilar BM (2008) Multi-level approach for the analysis of water effects in corn flakes. J Agric Food Chem 56(15):6447–6453. doi:10.1021/jf800541f

Grabowski S, Marcotte M (2003) Pretreatment efficiency in osmotic dehydration of cranberries. In: Welti-Chanes J, Velez-Ruiz JF (eds) Transport phenomena in food processing. Food preservation technology. CRC Press, Boca Raton. doi:10.1201/9781420006261.ch5

Haas GJ, Prescott HE, Cante CJ (1974) On rehydration and respiration of dry and partially dried vegetables. J Food Sci 39(4):681–684. doi:10.1111/j.1365-2621.1974.tb17956.x

Haddad J, Allaf K (2007) A study of the impact of instantaneous controlled pressure drop on the trypsin inhibitors of soybean. J Food Eng 79(1):353–357. doi:10.1016/j.jfoodeng.2006.01.066

Haddad J, Greiner R, Allaf K (2007) Effect of instantaneous controlled pressure drop on the phytate content of lupin. LWT—Food Sci Technol 40(3):448–453

Iguedjtal T, Louka N, Allaf K (2008) Sorption isotherms of potato slices dried and texturized by controlled sudden decompression. J Food Eng 85(2):180–190. doi:10.1016/j.jfoodeng.2007.06.028

Karathanos V (1993) Collapse of structure during drying of celery. Drying Technol 11(5):1005–1023. doi:10.1080/07373939308916880

Katz EE, Labuza TP (1981) Effect of water activity on the sensory crispness and mechanical deformation of snack food products. J Food Sci 46(2):403–409. doi:10.1111/j.1365-2621.1981.tb04871.x

Kaymak-Ertekin F, Gedik A (2004) Sorption isotherms and isosteric heat of sorption for grapes, apricots, apples and potatoes. LWT—Food Sci Technol 37(4):429–438. doi:10.1016/j.lwt.2003.10.012

Kerdpiboon S, Devahastin S, Kerr WL (2007) Comparative fractal characterization of physical changes of different food products during drying. J Food Eng 83(4):570–580. doi:10.1016/j.jfoodeng.2007.03.039

Kostaropoulos AE, Saravacos GD (1997) Thermal diffusivity of granular and porous foods at low moisture content. J Food Eng 33(1–2):101–109. doi:10.1016/S0260-8774(97)00036-8

Maritza AM, Sabah M, Anaberta CM, Montejano-Gaitán JG, Allaf K (2012) Comparative study of various drying processes at physical and chemical properties of strawberries (fragariavar-camarosa). Procedia Eng 42:297–312. doi:10.1016/j.proeng.2012.07.418

Mayor L, Sereno AM (2004) Modelling shrinkage during convective drying of food materials: a review. J Food Eng 61(3):373–386. doi:10.1016/s0260-8774(03)00144-4

Mazza G (1982) Moisture sorption isotherms of potato slices. Int J Food Sci Technol 17(1):47–54. doi:10.1111/j.1365-2621.1982.tb00158.x

Mounir S, Allaf K (2008) Three-stage spray drying: New process involving instant controlled pressure drop. Drying Technol 26(4):452–463. doi:10.1080/07373930801929334

Mounir S, Allaf K (2009) Study and modelling of dehydration and rehydration kinetics within porous medium. Association Française de Séchage pour l'Industrie et l'Agriculture, Lyon

Mounir S, Allaf K (2012) Enhancing hot air drying of chicken meat by instant controlled pressure drop DIC. Paper presented at the 1st Middle-East Drying Conference (MEDC2012), Mahshar, 19–20 Feb 2012

Mounir S, Besombes C, Al-Bitar N, Allaf K (2011a) Study of instant controlled pressure drop DIC—treatment in manufacturing snack and expanded granule powder of apple and onion. Drying Technol 29(3):331

Mounir S, Besombes C, Albitar N, Allaf K (2011b) Study of instant controlled pressure drop DIC treatment in manufacturing snack and expanded granule powder of apple and onion. Drying Technol 29(3):331–341. doi:10.1080/07373937.2010.491585

Mounir S, Halle D, Allaf K (2011c) Characterization of pure cheese snacks and expanded granule powders textured by the instant controlled pressure drop (DIC) process. Dairy Sci Technol 91(4):441–455. doi:10.1007/s13594-011-0023-8

Mounir S, Allaf T, Mujumdar AS, Allaf K (2012) Swell drying: coupling instant controlled pressure drop DIC to standard convection drying processes to intensify transfer phenomena and improve quality—an overview. Drying Technol 30(14):1508–1531. doi:10.1080/07373937.2012.693145

Mujumdar AS, Menon AS (1995) Drying of solids. In: Mujumdar AS (ed) Handbook of industrial drying, 2nd edn. Marcel Dekker, New York, pp 1–46

Pimpaporn P, Devahastin S, Chiewchan N (2007) Effects of combined pretreatments on drying kinetics and quality of potato chips undergoing low-pressure superheated steam drying. J Food Eng 81(2):318–329. doi:10.1016/j.jfoodeng.2006.11.009

Prachayawarakorn S, Tia W, Plyto N, Soponronnarit S (2008) Drying kinetics and quality attributes of low-fat banana slices dried at high temperature. J Food Eng 85(4):509–517. doi:10.1016/j.jfoodeng.2007.08.011

Rackis JJ, Wolf WJ, Baker EC (1986) Protease inhibitors in plant foods: content and inactivation. Adv Exp Med Biol 199:299–347

Rahman MS (2001) Toward prediction of porosity in foods during drying: a brief review. Drying Technol 19(1):1–13. doi:10.1081/drt-100001349

Rahman MS, Sablani SS (2003) Structural characteristics of freeze-dried abalone: porosimetry and puncture test. Food Bioprod Process 81(4):309–315

Rahman MS, Chen XD, Perera CO (1997) An improved thermal conductivity prediction model for fruits and vegetables as a function of temperature, water content and porosity. J Food Eng 31(2):163–170. doi:10.1016/s0260-8774(96)00060-x

Roos Y, Karel M (1991) Phase transitions of mixtures of amorphous polysaccharides and sugars. Biotechnol Prog 7(1):49–53. doi:10.1021/bp00007a008

Ruiz-López II, Ruiz-Espinosa H, Luna-Guevara ML, García-Alvarado MA (2011) Modeling and simulation of heat and mass transfer during drying of solids with hemispherical shell geometry. Comput Chem Eng 35(2):191–199. doi:10.1016/j.compchemeng.2010.05.007

Sandhu KS, Parhawk B (2002) Studies on the preparation of dehydrated potato cubes. J Food Sci Technol (Mysore) 39(6):594–602

Setyopratomo P, Fatmawati A, Sutrisna PD, Savitri E, Allaf K (2011) The dehydration kinetics, physical properties and nutritional content of banana textured by instantaneous controlled pressure drop. Asia-Pacific J Chem Eng 7:726–732. doi:10.1002/apj.624

Singh S, Raina CS, Bawa AS, Saxena DC (2006) Effect of pretreatments on drying and rehydration kinetics and color of sweet potato slices. Drying Technol 24(11):1487–1494. doi:10.1080/07373930600952834

Sulaiman I (2013) Approches fondamentales et expérimentales des processus d'autovaporisation et de modification structurelle de matériaux surchauffés pendant la détente instantanée contrôlée vers le vide. Université de la Rochelle, La Rochelle

Téllez-Pérez C (2013) Valorization of Mexican agriculture by preservation and drying by instant autovaporization; case of green pepper. University of La Rochelle, La Rochelle

Téllez-Pérez C, Sabah MM, Montejano-Gaitán JG, Sobolik V, Martínez CA, Allaf K (2012) Impact of instant controlled pressure drop treatment on dehydration and rehydration kinetics of green Moroccan pepper (Capsicum annuum). Procedia Eng 42:1077–1101. doi:10.1016/j.proeng.2012.07.491

Waldron KW, Smith AC, Parr AJ, Ng A, Parker ML (1997) New approaches to understanding and controlling cell separation in relation to fruit and vegetable texture. Trends Food Sci Technol 8(7):213–221. doi:10.1016/S0924-2244(97)01052-2

Whorton C, Reineccius GA (1995) Evaluation of the mechanisms associated with the release of encapsulated flavor materials from maltodextrin matrices. In: Encapsulation and controlled release of food ingredients, vol 590. ACS Symposium Series. American Chemical Society, pp 143–160. doi:10.1021/bk-1995-0590.ch013

DIC-Assisted Hot Air Drying of Post-harvest Paddy Rice

Sabah Mounir and Karim Allaf

Nomenclature

A_i and q_i	Coefficients of Crank solution according to the geometry of the sample
D_{eff}	Effective diffusivity of water within husks ($m^2\ s^{-1}$)
d_{eq}	Equivalent thickness of rice grain husks
k	Slope of $\ln((W_\infty - W)/(W_\infty - W_o))$ versus time (s^{-1})
P	DIC steam pressure (MPa)
t	Time (s)
t_d	DIC thermal treatment time (s)
t_D	Time taken to dry paddy rice from 32 to 12.5 % dry basis (min)
v_m	Absolute velocity of solid porous medium ($m\ s^{-1}$)
v_w	Absolute velocity of water flow within husks ($m\ s^{-1}$)
W_∞	Final water content, dry basis, in solid matrix $t \rightarrow \infty$
W	Water content, dry basis, in solid matrix at time t
W_i	Initial water content, dry basis, in solid matrix
W_o	Theoretical value of water content, dry basis, in solid matrix at starting time (% dry basis)
ρ_m	Apparent density of dry material ($kg\ m^{-3}$)
ρ_w	Apparent concentration of water in material ($kg\ m^{-3}$)
τ	Corresponds to Fick's number: $\tau = D_{\mathrm{eff}}\left(\frac{t}{d_p^2}\right)$

S. Mounir
Faculty of Agriculture, Department of Food Science, Zagazig University, Zagazig, Egypt

K. Allaf (✉)
Laboratory of Engineering Science for Environment LaSIE FRE 3474 CNRS, University of La Rochelle, 17042 La Rochelle Cedex 01, France
e-mail: kallaf@univ-lr.fr

T. Allaf and K. Allaf (eds.), *Instant Controlled Pressure Drop (D.I.C.) in Food Processing: From Fundamental to Industrial Applications*, Food Engineering Series, DOI 10.1007/978-1-4614-8669-5_2, © Springer Science+Business Media New York 2014

1 Introduction

Rice is one of the major cereals in the world economy. Paddy rice production stands at more than 730 million tons or 487 million tons of milled rice (FAO 2012). The moisture content of paddy rice at harvest can be as high as 28–35 % dry basis (db), especially when harvested during the rainy season. Paddy rice should be dried to typically 12–15 % db moisture content, which is known to be adequate for safe storage, milling, hulling, and whitening, and then stored as milled rice in order to prevent the production of mycotoxins (Hall 1970).

Standard drying usually involves moisture and temperature gradients within a kernel, inducing the development of a strong tensile stress gradient between the surface and the core of the kernel (Sharma and Kunze 1982). Different drying operations result in extensive kernel fissuring with a significant broken ratio during subsequent husking and milling. Hot air drying often results in over-dried grains, which become brittle and predisposed to cracking.

Drying and storage conditions (temperature, air moisture) affect all functional and structural properties (Pearce et al. 2001; Cogburn 1985). In an attempt to prevent kernels becoming more vulnerable to fissuring, researchers and the food industry have preferentially adopted low-temperature drying. Hence, the drying time is relatively long (from 8 to 12 or even 24 h) and involves a few tempering periods. Sarker et al. (1996) reported that a high initial moisture content or a high air temperature for drying resulted in a large number of fissured kernels. A related investigation indicated that the broken ratio could be related to the drying rate constant and the duration of drying (Chen et al. 1997). Others suggested the air temperature for drying should be kept below 53 °C to minimize the effect of thermal expansion on rice fissuring. Sharma and Kunze (1982) showed that only a small number of kernels fissured during drying and that most kernel fissuring occurred within 48 h after drying. After the drying stage, a milling/polishing process is carried out. Milling of the dried, cleaned paddy rice is to enable the following treatments:

- Hulling: to remove the external envelopes of the grain (hulls) to obtain brown rice
- Whitening: to eliminate the outer layers (pericarp) and the germ of the kernel by milling the brown rice to produce "white rice"

After sorting, white rice can be polished or glazed with a mixture of talc and glucose to enhance the product's commercial value and prolong its shelf life.

Paddy is the most suitable form for the storage of rice. Usually the storage life of unmilled paddy rice and brown rice is rather short, not exceeding a few months, and is limited by the presence of insects and larvae. Fundamentally, rice is subject to the same storage requirements as any other cereal: it should be kept in cool, dry, dark conditions. Milling, which is performed to make white rice, removes the outside silver skin (aleurone layer) and the rich-in-nutrients grain germ. This increases the shelf life of rice, but at the cost of losing a large proportion of its nutrients. Furthermore, since rice is likely to absorb odors very quickly, it should be kept well away from sources of strong smells. Depending on the variety, when stored in a dry, tightly sealed container, white rice can be stored for up to 3 years.

Despite taking precautions and using low drying kinetics, the broken rice ratio is still too high and results in a relatively low yield of head rice. After industrial milling, 100 kg of paddy rice yields about 60 kg of white rice, 10 kg of broken grains, 10 kg of bran and flour, and 20 kg of hulls (FAO 2012).

The cooking time for both brown and white rice is long, close to 50 and 17 min, respectively, and only 4 min later it is overcooked.

With paddy rice, the instant controlled pressure drop (DIC) process is employed as a post-harvesting treatment to reduce drying time, improve yield, and produce higher quality brown and white rice (short cooking time, etc.).

Indeed, in numerous new unit operations, DIC technology has greatly intensified the limiting transfer phenomenon, reduced energy consumption, and can be considered an environmentally friendly process (Al Haddad 2007). DIC treatment is followed by hot air drying and shade polishing.

An optimized DIC treatment can improve the multidimensional qualities of the product in terms of its nutritional, hygienic, organoleptic, and convenience attributes. DIC treatment has many advantages in terms of processing time (not exceeding 30 s whatever the variety) and drying kinetics (about 2–3 h instead of 1 day under standard 45–50 °C hot airflow conditions). Shade polishing yields up to 67–70 % of high-quality whole-grain rice rather than the usual 52–60 %. A tasting evaluation was carried out by an international panel and the rice was found to be perfect after cooking for 6 min, and overcooked after 18 min, compared to 17 and 20 min, respectively, for the same variety of standard products. The shaded shelf life of the final product far exceeds 2 years.

2 Materials and Methods

2.1 Raw Material

A large panel of Vietnamese paddy rice was selected for investigation, with different initial moisture contents measured just after harvesting (from 28 to 35 % db), and diverse sizes and forms; with 15–25 % amylase. The samples of raw paddy rice were sorted in order to completely eliminate extraneous bodies as well as green grains and empty husks.

2.2 Treatment

In the present study a simple sorting operation was performed just after harvesting to remove waste from the paddy rice. DIC treatment is normally included as a post-harvesting stage, followed by hot air drying. The treatment design is illustrated in Fig. 1, which is a flow diagram of the main operations carried out on the paddy rice just after or 6–8 h after harvesting.

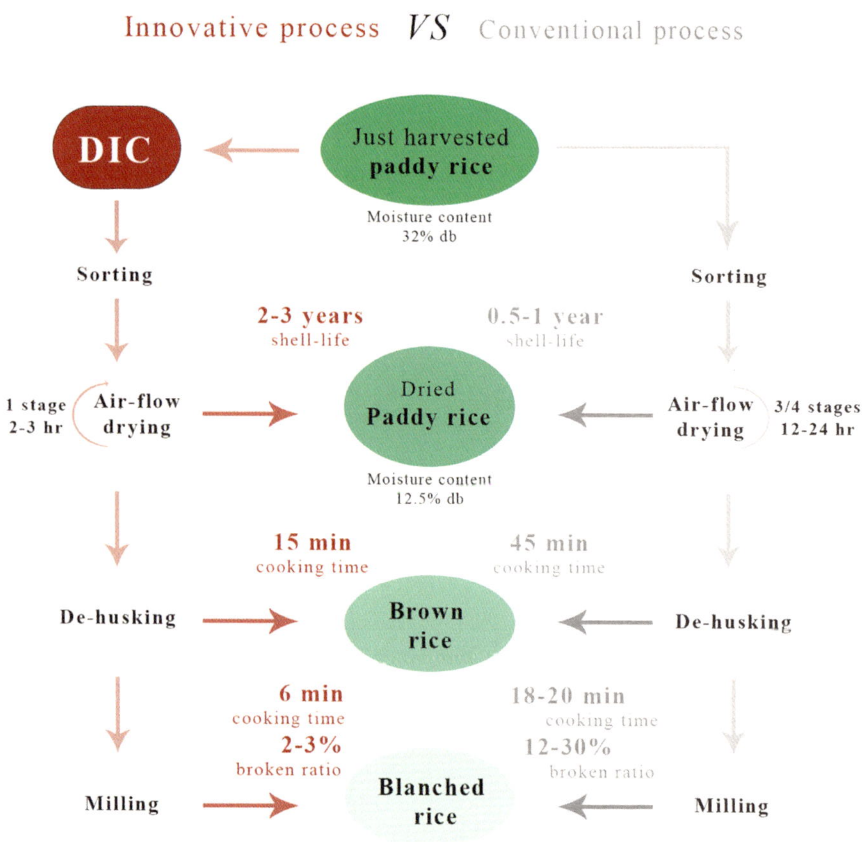

Fig. 1 Rice treatment protocol of post-harvest DIC-assisted drying

2.2.1 The Instant Controlled Pressure Drop (DIC) Process

The DIC process has been defined as two main stages of: (1) thermal treatment carried out under a high saturated steam pressure of 0.1–0.6 MPa, which allows the temperature of the product to attain between 100 and 160 °C, for a short duration of 5–60 s followed by: (2) an abrupt pressure drop towards a vacuum (4–5 kPa) with a very high decompression rate (higher than 0.5 MPa s^{-1}) which is generally maintained for 5–20 s. During the first stage, steam condensation on the surface of the grain helps to heat the paddy rice. This superficial moisture requires time to diffuse down to the core and reach homogeneous moisture and temperature levels before the pressure drops.

2.2.2 Drying Process

As soon as the paddy rice is recovered from the DIC reactor it is dried. The purpose of the drying stage is to reach a water content close to 12–14 % db. A static bed drier is used with a 2 cm thin layer of DIC-treated or -untreated paddy rice which is placed under an airflow at a constant input temperature of 50 °C. The speed of the airflow should be higher than 1.25 m s^{-1}.

Pilatowski et al. (2010) assumed the transfer mechanisms to be external and internal heat and mass transfers in this type of thermal drying process. Heat is generally transferred from air to the surface of the grain by convection and from the surface towards the grain core by conduction. There is mass transfer within the product, which, in the present case, is assumed to occur only through the liquid. Different phenomena (capillary flow, diffusion, etc.) may occur; however, the moisture gradient remains the driving force.

Duong Thai (2003) proved that mass transfer was the limiting transfer process within a paddy rice husk since it had the highest resistance to moisture diffusion (Duong Thai et al. 2008). Indeed, heat transfer and the homogenization of temperature distribution within the grain must be conducted more quickly because of the higher diffusivity of heat compared to moisture. Thus, the entire process of heating by condensation of saturated steam is controlled by the transfer of moisture within the grain; numerous studies and experiments have already confirmed that the mass diffusivity D_{eff} of moisture in a porous material is much lower than the thermal conduction diffusivity α:

$$D_{\mathrm{eff}} \ll \alpha \tag{1}$$

Thus, by assuming drying *kinetics* is controlled by the mass transfer of water within the husks, Mounir and Allaf (2009) proposed and adopted the air-washing/diffusion model. This model is used to obtain an initial airflow that washes the surface, followed by a Fick-type diffusion process.

Experimental values can be used in the actual diffusion model to identify the effective diffusivity D_{eff}. It should be noted that the experimental data relative to the starting period should not concern the diffusion process. Furthermore, values related to the final paradoxical stage of drying (Mounir and Allaf 2008) must not be included in the model. The theoretical value W_{o}, calculated by extrapolating the diffusion model at $t = 0$, should be distinct from the real value of the initial water content W_{i}. The difference ($\delta W_{\mathrm{s}} = W_{\mathrm{i}} - W_{\mathrm{o}}$) between this theoretical value, W_{o}, and the real one W_{i} is the starting accessibility (δW_{s}), with corresponds to the amount of water extracted at the beginning through the rapid interaction between the airflow and the surface, independently of any diffusion process. By modifying the matrix structure, the values of D_{eff}, W_{∞}, and δW_{s} vary and may be considered as the main response parameters characterizing the functional impact of DIC treatment.

2.3 Assessment Protocol

Paddy rice grains with a water content close to 12.5 % dry basis were dehusked and polished. The resulting white rice was separated into two categories: unbroken grains (more than nine-tenths the length of normal head rice) and broken grains. The yield and the broken ratio were consequently determined according to the following equations:

$$\text{Yield of unbroken grain} = \frac{\text{Mass of unbroken grain}}{\text{Total mass of white rice}} \tag{2}$$

$$\text{Broken ratio} = \frac{\text{Mass of broken grain}}{\text{Total mass of white rice}} \tag{3}$$

Cooking time can be determined using sensorial or instrumental methods. The first method was defined by Habba (1997). The second tasting method was conducted by a six-person international panel (one Malaysia, one Middle East, one North Africa, two France, and one Mexico). The two responses selected were the time required to reach a good level of cooking and the time required just to reach the overcooked state.

3 Results and Discussion

3.1 Drying Kinetics

Pilatowski et al. (2010) studied the kinetics of airflow drying at 50 °C with DIC-treated Mexican paddy rice just after harvesting (Morelos-A98). Here, too, the drying time t_D needed to reach 12.5 % dry basis from an initial moisture content of 32 %, was determined. DIC treatment significantly reduced the drying time of Paddy rice t_D by a factor of 5.5, down from 1,110 to 205 min.

The second main objective was to estimate the effective diffusivity D_{eff} of moisture within the grain husks. The results were systematically in agreement with the surface airflow washing/diffusion model with R^2 between 0.912 and 0.995. From a scanning electron microscope (SEM) image (Fig. 2), it was possible to calculate the equivalent thickness d_{eq} of husks from ten repeated measurements; the mean value was 79.4 ± 0.5 μm.

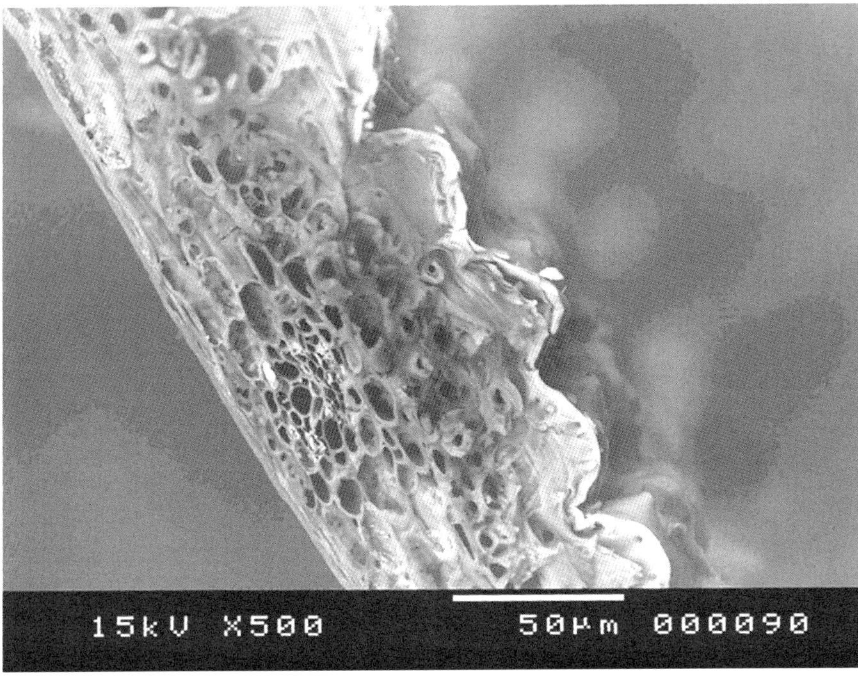

Fig. 2 SEM of husk of Mexican paddy rice (Morelos-A98)

The optimum response in terms of maximizing the effective diffusivity D_{eff} was calculated to be $D_{eff} = 1.18 \pm 0.06$ E−13 $m^2\ s^{-1}$ for DIC treatment at a steam pressure of 0.54 MPa and a processing time of 16 s; experimental trials confirmed this value compared with $D_{eff} = 0.22 \times 10^{-13}\ m^2\ s^{-1}$ for untreated paddy rice.

3.2 Hulling, Polishing, and Yields

Breaking mainly occurs during the drying process as a result of the humidity gradient, causing mechanical strains within the grain. In standard processes, in spite of tempering periods during drying to minimize the amount of breakage, rice has a high broken ratio, exceeding 21 % of the blanched rice. Conversely, DIC treatment achieved before hot air drying, dehusking, and subsequently polishing in order to obtain white rice; resulted in a great amount of head rice. The yield that was attained in this study was around 69.5 % of the initial total amount of paddy rice. Thus, DIC treatment considerably reduces the percentage of broken grain ratio to between 1.5 and 3 %.

3.3 Cooking Time

With an international panel formed specifically for the purpose of carrying out a tasting test, the minimum cooking time and the starting point for overcooking were determined to be 5 and 18 min, respectively, for DIC-treated samples compared to 16 and 21 min, respectively, for untreated samples.

Another aspect of "cooking" time was calculated by immersing rice in cold water (20 °C). In less than 2–3 h, white rice was perfectly cooked with completely gelatinized starch.

4 Energy Consumption, Environmental Impact, and Cost

Shade experiments were used to calculate the total energy consumption of DIC treatment, which was about 250 kWh/ton of paddy rice. On a pilot plant scale, drying paddy rice from 30.5 to 12.5 % dry basis required 215 kWh/ton after DIC treatment compared with 750 kWh/ton with untreated paddy rice (Qinghua et al. 2002). From these results we concluded that the energy demand for both DIC treatment and final hot air drying was lower (415 kWh/ton instead of 750 kWh/ton) in relation to standard hot air drying, independently of operational handling.

The increase in storage time to 18 months instead of the usual 9 months, together with yields from whole grains of 97–98 % of head rice compared to 79 % for untreated rice, would confer a significant economic advantage on DIC-assisted drying if it were used just after harvesting.

The potential energy of a husk is about 4,200 kWh/ton of husk, which means that 1 ton of paddy rice contains about 840 kWh husk energy. Indeed, during the milling of paddy rice, about 70 % of the weight is recovered as broken and unbroken white grain rice. Bran represents about 8 % of the weight while 22 % of paddy rice is recovered as husk. This husk can advantageously be used as fuel in the rice mills to generate steam as well as electricity for parboiling. Paddy husk contains about 75 % volatile organic matter, and 1 ton equivalent husk has an energy equivalent to 0.75 TEP (ton equivalent petrol). The final rice husk ash is about 25 % of the weight of the husk, with around 85–90 % of amorphous silica.

So 1,260 tons of just harvested paddy rice at 30 % wb (wet basis) water content can be dried using DIC post-harvest treatment to reach the required value of 12 % wb in less than 2–3 h, without any tempering period; 1,000 tons of adequately dried paddy rice can then be obtained. The energy needed for the total operation of DIC-assisted drying to remove about 260 tons of water would be around 465 MWh. 700 tons of unbroken white grain rice and about 220 tons (22 %) of husk would be produced. By burning this husk, about 140 TEP or about 1,500 MWh could be generated.

The USDA report says that Egypt's paddy rice production in the year May 2012 to April 2013 is expected to rise to 6.37 million tons of dried paddy rice, i.e., 4.5 million tons of dried unbroken white grain DIC rice. This would be equivalent to 1.4 million tons of husk, corresponding to about 0.5 million TEP or 2.5 million barrels of petrol. DIC-assisted drying would need less than 33 % of this energy to process paddy rice.

5 Conclusion

The technology using instant controlled pressure drop (DIC) as a post-harvesting treatment for paddy rice, followed by conventional airflow drying, is one of the most efficient drying processes. The most appropriate way to use this type of two-stage drying is to apply DIC treatment just after or 6–9 h post-harvest; standard airflow drying is then very efficient. This drying strategy enables the user to increase drying performance, reduce costs, and optimize product quality. The impact in terms of decreasing the drying time is impressive: 205 min compared with 1,110 min and no tempering period needed. DIC was subsequently optimized and carried out under a steam pressure of 0.5 MPa for 16–30 s. Minimizing the broken ratio is particularly important and it was reduced to around 3 % compared to 21 % for untreated rice. The big advantage of using DIC is that the rice preserves its aromatic properties for longer. Regarding cooking, DIC rice is characterized by a short cooking time compared with untreated rice (5 min versus 15 min) and a lengthy period before the overcooked point is reached (about 18 min as against 20 min). Finally, it is worth noting the impact of DIC treatment in terms of microbial decontamination and elimination of insects, which allows a long shelf life, in excess of 18 months.

According to many studies carried out in university laboratories and in industry, the main advantages of this two-stage DIC drying have been established for a large range of paddy rice varieties.

DIC treatment used just after harvesting/sorting and before any drying stage removes the weaknesses of "normal processing" by:

- Reducing drying time (2–3 h instead of 12–24 h)
- Increasing the whole DIC drying performance through more homogeneity and lower energy consumption
- Increasing industrial processing capacities
- Reducing the broken ratio (1–3 % instead of 10–30 %), thereby increasing yields (by about 10 %)
- Shortening cooking time (6 min instead of 17 min) with a longer time before overcooking (reached 16 min after cooking instead of 4 min)

(continued)

- Preserving organoleptic content and improving grain quality
- Maintaining aromatic content
- Improving intrinsic nutritional value
- Increasing shelf life (2–3 years instead of 6 months to 1 year)

The use of husks as an energy source would be very important. It could ensure about three times more than the total energy needed for DIC-assisted drying and processing of paddy rice. In the case of Egypt, the annual rice production is about 6.4 million tons of paddy rice. Using DIC-assisted drying, the annual production of unbroken white grain rice could reach 4.5 million tons (compared with of 3.6 million tons) and 1.4 million tons of husk. This corresponds to 2.5 million barrels of petrol; only 33 % of this energy would be used in the DIC drying process as such. It would be interesting to explore the use of the bran (0.5 million tons) and ash (0.3 million tons) end products for nutritional purposes and as high-quality soil fertilizer, respectively.

References

Al Haddad M (2007) Contribution théorique et modélisation des phénomènes instantanée dans les opérations d'autovaporisation et de déshydratation. Université de La Rochelle, La Rochelle

Chen H, Siebenmorgen TJ, Marks B (1997) Relating drying rate content to head rice yield reduction of long grain rice. Trans ASAE 40(4):1133–1139

Cogburn RR (1985) Rough rice storage. In: Juliano BO (ed) Rice: chemistry and technology. St. Paul, American Association of Cereal Chemists, pp 265–287

Duong Thai C (2003) Etude de l'application du procédé hydro-thermique dans le traitement de différents types de riz: procédé d'étuvage et micro-expansion par détente instantanée contrôlée et impact sur les propriétés fonctionnelles (Study of hydro-thermal treatment of various varieties of paddy rice: steaming process and micro-expansion using instant controlled process; impacts in terms of functional properties). Université de La Rochelle, La Rochelle, France

Duong Thai C, Al Haddad M, Rezzoug Z, Lefevre L, Allaf K (2008) Dehydration by successive pressure drops for drying paddy rice treated by instant controlled pressure drop. Drying Technol 26(4–6):443–451

FAO (2012) Rice market monitor, vol XVI. Food and Agriculture Organization, Rome

Habba A (1997) Nouveaux procédés de précuisson et d'étuvage du riz par détente instantanée contrôlée (DIC): analyse des opérations, optimisation des procédés et approche à l'industrialisation. Université de Technologie de Compiègne, Compiègne

Hall DW (1970) Handling and storage of food grains in tropical and sub-tropical areas, vol 90. FAO Agriculture Development, Rome

Mounir S, Allaf K (2008) Three-stage spray drying: new process involving instant controlled pressure drop. Drying Technol 26(4):452–463. doi:10.1080/07373930801929334

Mounir S, Allaf K (2009) Study and modelling of dehydration and rehydration kinetics within porous medium. Association Française de Séchage pour l'Industrie et l'Agriculture, Lyon

Pearce MD, Marks BP, Meullenet J-F (2001) Effects of postharvest parameters on functional changes during rough rice storage. Cereal Chem 78(3):354–357

Pilatowski I, Mounir S, Haddad J, Cong D, Allaf K (2010) The instant controlled pressure drop process as a new post-harvesting treatment of paddy rice: impacts on drying kinetics and end product attributes. Food Bioprocess Technol 3(6):901–907. doi:10.1007/s11947-010-0332-6

Qinghua Z, Yang SX, Mittal GS, Shujuan YI (2002) Prediction of performance indices and optimal parameters of rough rice drying using neural networks. Biosyst Eng 83(3):281–290

Sarker NN, Kunze OR, Strouboulis T (1996) Transient moisture gradients in rough rice mapped with finite element model and related to fissures after heated air drying. Trans ASAE 39 (2):625–631

Sharma AD, Kunze OR (1982) Post-drying fissure developments in rough. Trans ASAE 25:465–468

DIC-Assisted Parboiling of Paddy Rice

Cong Duong Thai and Karim Allaf

1 Introduction

Storage of paddy rice generally cannot exceed a limited period of a few months, even after drying, which reduces moisture content close to 12 % db (dry basis). Drying and storage temperature conditions affect all functional properties (Pearce et al. 2001; Cogburn 1985).

Before the hulling and polishing steps Paddy rice can be parboiled, a thermal treatment using steam that gives rise to low bonding rice kernels. This is a specific method that involves soaking, where the humidity goes from 12 % dry basis (db) to about 30 % db; steam treatment, during which pressure is applied to ensure the water-soluble nutrients, vitamins, and minerals that are initially contained in the aleurone layer migrate towards the grain kernel; and drying. Parboiled rice obtained in this way has a greater nutritional content due to the partial migration of vitamins and minerals to the endosperm (CFTRI 1969; Bhattacharya and Subba Roa 1966).

2 Parboiling: Issues and Intensification

2.1 Traditional Parboiling: Stages and Issues

Parboiling consists of soaking, pressure steaming, and drying steps prior to milling. Paddy rice is the principal raw material used in the production of parboiled rice (Bhattacharya 1985). Parboiling gelatinizes the starch granules and hardens the

C.D. Thai
CanTho University of Technology, CanTho City, Vietnam

K. Allaf (✉)
Laboratory of Engineering Science for Environment (LaSIE FRE 3474) CNRS, 17042 La Rochelle Cedex 01, University of La Rochelle, La Rochelle, France
e-mail: kallaf@univ-lr.fr

T. Allaf and K. Allaf (eds.), *Instant Controlled Pressure Drop (D.I.C.) in Food Processing: From Fundamental to Industrial Applications*, Food Engineering Series, DOI 10.1007/978-1-4614-8669-5_3,

endosperm, making it translucent. A white core indicates that the parboiling process is not yet complete. Polishing the rice becomes easier and it takes less time to cook than white rice. The parboiled rice is then cooled and dried before storage and milling. White parboiled rice has higher nutritional value, thanks to the migration and retention of natural vitamins and minerals in the bran kernels. The rice is usually slightly yellowish, although the color largely fades after cooking (Pillaiyar and Mohandoss 1981).

Traditional parboiling begins with a 2–4 h soaking stage, although at a lower temperature than for gelatinization. Once the required water content of the paddy rice has been reached, a 1–2 h thermal treatment is carried out using dry steam (Soponronnarit et al. 2006). Conventional parboiling of 1 ton of paddy rice generally requires an average volume of 800–1,000 L of water for the soaking operation. Around 180–220 L of water is absorbed by the rice. Recycling the residual water (~800 L) is difficult because of the presence of impurities, including many organic substances such as starch and mineral salts.

The starch in the rice kernel is then gelatinized, generally by steaming at 100–120 °C. It is therefore necessary to dry the steamed rice. This needs to be done very carefully in order to avoid breaking the rice kernels. In the industrial process the steamed paddy rice, whose moisture content is nearly 22 %, is dried under airflow. After the first tempering period, the paddy rice is dried to a moisture content of 16–18 % using hot air, followed by a second tempering period. Then, there is a final drying stage to reach a humidity of about 12–13 % db, which prepares the rice for hulling and polishing. Thus, the airflow drying time is relatively long (from 8 to 12 h) and requires two or three 4-h periods of rest (tempering time), which aims at establishing a homogeneous distribution of moisture throughout the grain. However, despite these precautions, the percentage of broken rice is still generally above 9 % (and can even reach 25 %) and results in a relatively low yield of unbroken white parboiled rice. Moreover the cooking time is often close to 18–20 min. Finally it should be noted that the conventional steaming of aromatic rice is not possible because the majority of the flavor content is removed (Ali and Ojha 1977).

2.2 DIC Steaming Process

2.2.1 Soaking Process

Instant controlled pressure drop (DIC) parboiling of rice is generally carried out with dried paddy rice. As with conventional parboiling, DIC steaming needs a soaking stage to increase the grain moisture to the optimal value for starch gelatinization (30–40 % db). The optimization of this process leads to a quick and uniform absorption of water. Generally, the lower the water temperature, the slower the soaking process. However, the temperature should not exceed the gelatinization temperature. When water is used at ambient temperature (20–30 °C), soaking the rice takes 36–48 h. This long soaking period may dissolve some of the rice in the water; seeds begin to germinate; and starch fermentation occurs. In hot water (60–65 °C), soaking only takes 2–4 h, depending on the variety (Indudhara Swamy et al. 1971). Soaking time can be reduced to 60–90 min by first exposing paddy rice to a vacuum for a few minutes and/or by soaking it under pressure in hot water.

2.2.2 Parboiling Using Instantaneous Controlled Pressure Drop

After soaking the already dried paddy rice, which is generally performed under a vacuum at 65 °C, it is parboiled using DIC. DIC-assisted parboiling is a short-time (25–30 s) high-pressure (up to 0.6 MPa) saturated steam thermal treatment (Duong Thai et al. 2008) followed by an instant ($\Delta P/\Delta t > 5$ MPa s^{-1}) pressure drop towards a vacuum (5 kPa). The aim of the later stage is to dry the parboiled paddy rice to reach a level close to 11 % db moisture content. Thanks to DIC treatment, the drying is achieved in 2–3 h without any need to insert a tempering period. It adequately prepares paddy rice for good hulling and polishing with a very low percentage of broken kernels (generally less than 3 %), and thus yields a very high quantity of unbroken white rice.

It is easy to control the cooking time for DIC-parboiled rice, which is between 7 and 15 min and mainly depends on the rate of the pressure drop. Perfectly optimized DIC-assisted parboiling could be used with many varieties of rice to obtain a very high yield (up to 66–70 g of whole white grain/100 g of paddy rice). Furthermore, it can be applied to aromatic varieties of rice since it preserves their aromatic content.

2.3 *Rice Varieties*

Numerous varieties of rice have been studied: Morelos-A98 (Mexican rice), 28 Vietnamese varieties, two Thai varieties, and the Giza 180 Egyptian variety; they can be distinguished by their form and their composition (amylose–amylopectin content) (Salcedo-Aceves 2006). The harvested paddy rice had an initial water content of between 28 and 35 % db.

An assessment of the raw material was made in terms of physicochemical characteristics such as amylose–amylopectin content, gelatinization temperature range, average size of the grains among the different varieties used in our studies, as well as cooking time. Some of these results are shown in Table 1. The amylose content of these varieties ranged between 16.15 and 27.51 %, and amylopectin content between 71.12 and 84.10 %. The gelatinization temperature alternated between low, intermediate, and high.

Scanning electron microscopy (SEM) was used to identify the structure, focusing on the hulls and bran of paddy rice. The internal alveolated structure of the paddy rice hulls could explain the functional behavior of this rice in terms of the soaking,

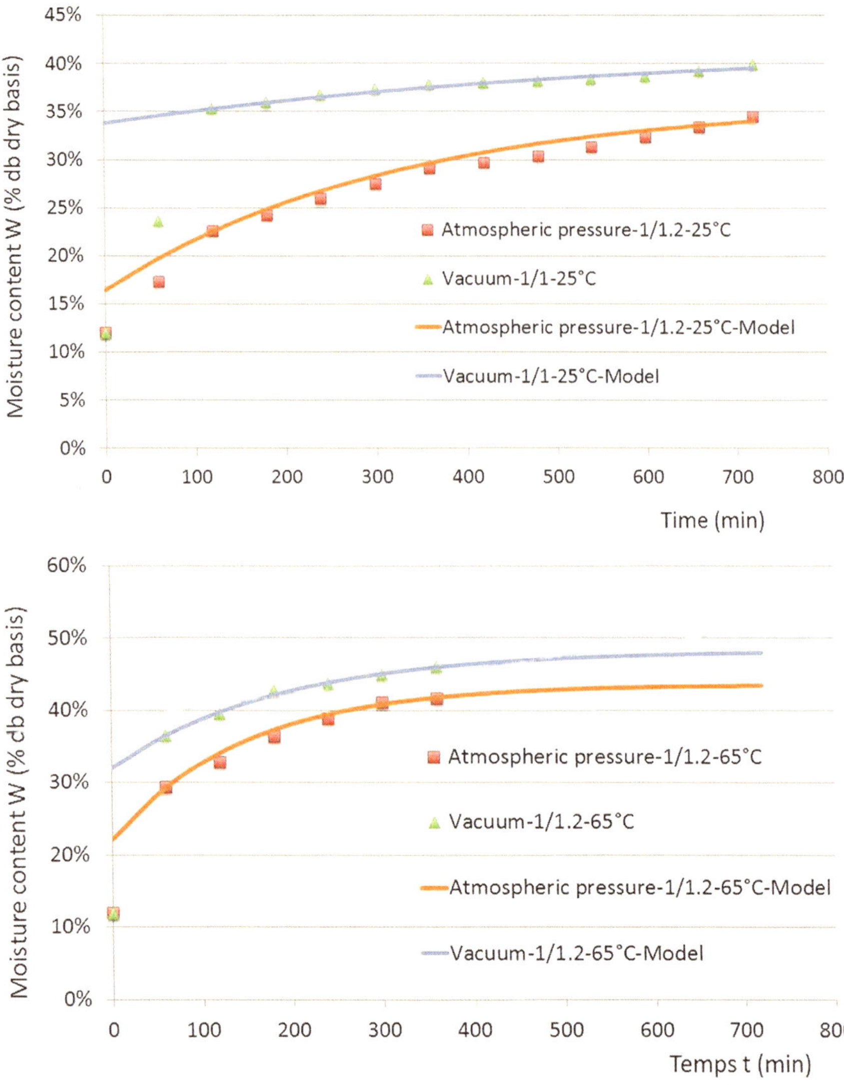

Fig. 1 Soaking kinetics of rice in water at 25 and 65 °C, with a water/rice ratio of 1.0 and 1.2 under normal pressure (*N*) and with an initial vacuum stage (*V*)

720 min at 25 °C, and in 60 min compared with 180 min at 65 °C, with an initial vacuum stage and at atmospheric pressure.

The model of superficial washing/internal water diffusion (Mounir and Allaf 2009) was applied with R^2 between 0.91 and 0.997. The starting accessibility δW_s seems to be independent of temperature and greatly dependent on pressure:

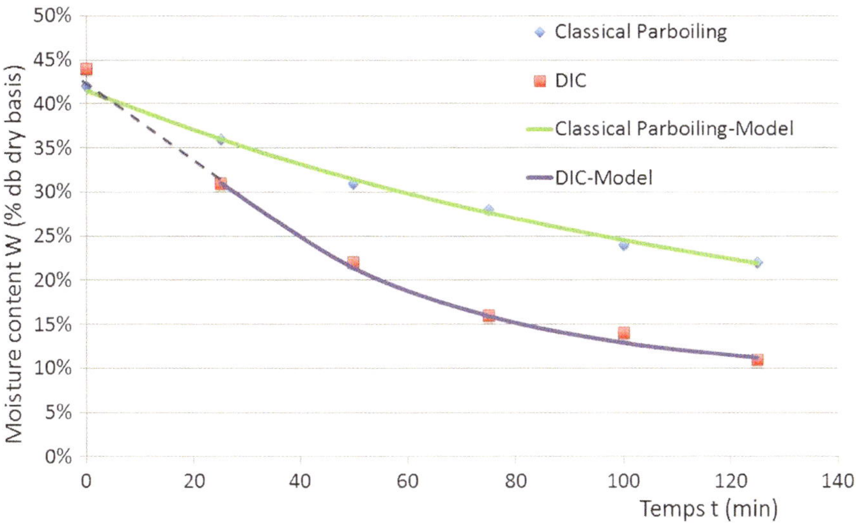

Fig. 2 Kinetics of drying by hot airflow at 50 °C of standard and DIC-parboiled rice samples

13 and 23 % db for soaking carried out under atmospheric pressure and with an initial vacuum, respectively. The water diffusivities D_{eff} seemed to be independent of pressure, but strictly linked to temperature: 2–3 × 10^{-8} m^2/s and 1.3 × 10^{-7} m^2/s, for 25 °C and 65 °C, respectively.

3.2 *Drying Kinetics*

Hot air drying was systematically carried out after the actual parboiling treatment. The aim was to lower water content from an initial level of about 42 % db. An experimental design was devised to optimize the treatment. So, drying with a hot airflow at 50 °C, 1.5 m/s, and 265 Pa partial pressure of humidity was carried out with Mexican paddy rice after the different DIC treatments according to the experimental design. The drying kinetics was compared with samples previously treated by standard parboiling. The main objective was to estimate the time required to reach a moisture content of 12.5 % db, which is the right amount for good storage and the subsequent hulling and polishing processes.

Figure 2 shows the change in water content dry basis versus drying time.

The response parameters (dependent variables) were drying time and effective diffusivity (D_{eff}), calculated using Mounir and Allaf's (2009) model which is based on an initial air washing followed by a stage limited and controlled by the internal diffusion of the water within the hull.

DIC parboiling significantly reduced the drying time of samples and increased effective diffusivity with Mexican rice, to 201 min and 9.87×10^{-11} m^2/s compared with 1,110 min and 3.0×10^{-11} m^2/s, respectively.

The optimized operating parameters maximizing effective diffusivity D_{eff} and minimizing drying time were calculated to be a saturated steam pressure of 0.54 MPa and a treatment time of 16 s.

3.3 Hulling, Polishing, and Performance

After DIC parboiling and hot air drying, the samples of paddy rice were hulled and polished to obtain white rice, with a performance in this case estimated to be 69.5 % of the original paddy rice. With this process it is possible to separate and count the number of whole grains and broken grains. A grain is considered broken when it is less than 70 % of the length of a normal grain. Breaking is mainly generated during the drying process because of the high moisture gradient generated in the grain. Despite the two or three tempering periods during the drying process in standard parboiling, the Mexican rice had a large number of broken kernels, more than 23 % of the milled rice. With DIC parboiling it was observed that the number of broken rice grains was as low as 1.5–4.5 % of milled rice, despite the fact that drying was carried out without any tempering period.

3.4 Cooking

Sensory analysis was conducted by a specific international panel who determined the minimum cooking time and the time at which the rice was overcooked. A good texture of cooked rice was scored between −1 and +1. For uncooked hard grain samples, the scores were from −5 to −1, and between +1 and +5 for overcooked samples. Figure 3 shows another major advantage of DIC-parboiled rice (similar to the samples treated by DIC just after harvesting): instant cooking, within 5 min compared with 15 min for non-DIC rice, and a relatively long time before it was overcooked (13 min compared with 5 min for non-DIC rice).

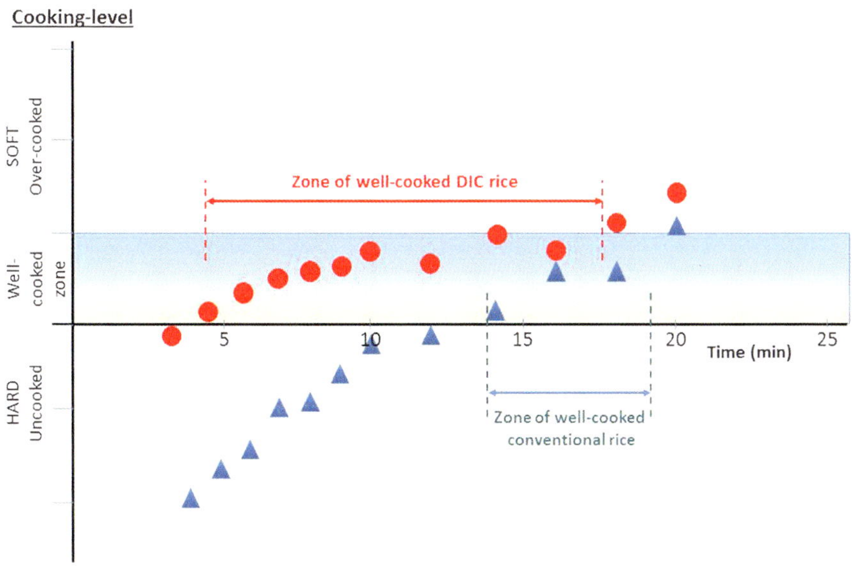

Fig. 3 Change in textural state of grains versus the immersion time in boiling water

White DIC-parboiled rice was well cooked very rapidly (5 min) and maintained a good texture, without being overcooked at 18 min, compared with 15 min and 20 min, respectively, for standard parboiled rice.

4 Conclusion

A specific DIC-assisted parboiling treatment for different varieties of paddy rice had the same impact as post-harvesting DIC treatment in terms of final product quality and process performance, reducing drying time (205 min compared to 1,110 min). DIC treatment was optimized at 0.5 MPa saturated steam pressure and 16–30 s thermal treatment. The broken kernel ratio fell substantially, at less than 3 % compared to 25 % with standard treatments. The rice obtained was characterized by the fact that the aromatic content was well preserved. DIC-treated rice is distinguished by both a low cooking time (5 min compared to 15 min) and a long period before the rice is overcooked (about 18 min compared to 20 min). Finally, the impact of DIC treatment in terms of microbial and insect decontamination is worth noting, with a storage time exceeding 18 months.

References

Ali N, Ojha TP (1977) Coukking quality of raw rice and paraboiled rice. Rice Process Eng Centre Rep 3(1):22–26

Bhattacharya KR (1985) Rough rice storage. In: Juliano BO (ed) Rice: chemistry and technology. American Association of Cereal Chemists, St. Paul, pp 289–348

Bhattacharya KR, Subba Roa PV (1966) Effect of processing conditions on quality of parboiled rice. J Agric Food Chem 14(5):476–479. doi:10.1021/jf60147a009

CFTRI (1969) Parboiling of paddy, vol Project Circular 7. Central Food Technological Research Institute, Mysore

Cogburn RR (1985) Rough rice storage In: Juliano BO (ed) Rice: Chemistry and Technology. American Association of Cereal Chemists, USA, pp 265–287

Duong Thai C, Al Haddad M, Rezzoug Z, Lefevre L, Allaf K (2008) Dehydration by successive pressure drops for drying paddy rice treated by instant controlled pressure drop. Drying Technol 26(4–6):443–451

Indudhara Swamy YM, Zakiuddin Ali S, Bhattacharya KR (1971) Hydration of raw and parboiled rice and paddy at room temperature. Sci Technol 8(1):20–22

Mounir S, Allaf K (2009) Study and modelling of dehydration and rehydration kinetics within porous medium. Lyon, France

Pearce MD, Marks BP, Meullenet J-F (2001) Effects of postharvest parameters on functional changes during rough rice storage. Cereal Chem 78(3):354–357

Pillaiyar P, Mohandoss R (1981) Hardness and colour in parboiled rices produced at low and high temperatures. J Food Sci Technol, 18:7–9

Salcedo-Aceves J (2006) Morelos A-98 variedad de arroz para Morelos y otros estados de la republica, Morelos (Mexico), vol. 21. INIFAP

Soponronnarit S, Nathakaranakule A, Jirajindalert A, Taechapairoj C (2006) Parboiling brown rice using super heated steam fluidization technique. J Food Eng 75 (3):423–432. doi: http://dx.doi.org/10.1016/j.jfoodeng.2005.04.058

Part II
Allergens and Decontamination

Effect of DIC on the Allergenicity of Legume Proteins

Carmen Burbano and Carmen Cuadrado

1 Introduction

In the past few decades, the prevalence of allergic diseases has increased dramatically and food allergies play a major role in this increase. Food allergy affects approximately 6–8 % children and 3–4 % adults in Europe (Mills et al. 2007). Legumes are increasingly regarded as beneficial food ingredients. In fact, they are recommended as a staple food by health organizations, and dieticians now tend to encourage their consumption in counseling (Leterme 2002). The allergenic potential of many legumes such as lupines (*Lupinus albus*), peanuts (*Arachis hypogaea*), lentils (*Lens culinaris*), chickpeas (*Cicer arietinum*), and soybeans (*Glycine max*) is well established (Verma et al. 2012). As legume allergy affects a significant proportion of the population, some effective methods should be adopted to minimize its risk. Consumption of legumes may provoke mild to severe anaphylactic symptoms in sensitized individuals. The usual symptoms are angioedema, vomiting, urticaria, allergic rhinitis, diarrhea, skin rashes, swelling of the tongue or throat, and asthma. Moreover, there is a significant degree of immunological cross-reactivity within the group of legume allergens and between this group and other plant allergens. Consequently, there is a growing interest in the development of newer methods to overcome this type of allergic problem and several strategies are being tested to minimize the allergenicity potential of legume crops (Chung and Reed 2011).

Heating promotes protein denaturation, aggregation, and structure disruption and it can therefore modify the allergenic properties of proteins. The molecular basis of changes in allergenic activity is the inactivation or destruction of epitope structures, the formation of new epitopes, or an enhanced access to cryptic epitopes by denaturation of the native allergen (Besler et al. 2001). Plant protein

C. Burbano (✉) • C. Cuadrado
Departamento de Tecnología de Alimentos, SGIT-INIA, Ctra. Coruña Km 7.5, 28040 Madrid, Spain
e-mail: burbano@inia.es

T. Allaf and K. Allaf (eds.), *Instant Controlled Pressure Drop (D.I.C.) in Food Processing: From Fundamental to Industrial Applications*, Food Engineering Series,
DOI 10.1007/978-1-4614-8669-5_4,

allergenicity may be variably affected by thermal processing due to an increase or decrease in IgE immunoreactivity. Thus, the overall effect of such procedures on a complex food matrix cannot be predicted (Beyer et al. 2001; Mondoulet et al. 2005). A better understanding of how thermal processing induces biochemical and immunological changes in food allergens may contribute to the development of new diagnostic tools and to alleviate the problem of allergies.

Our previous studies have shown that boiling in an autoclave under harsh conditions markedly reduced lupine, lentil, chickpea, soybean, and peanut allergenicity (Álvarez-Álvarez et al. 2005; Cabanillas et al. 2012; Cuadrado et al. 2007, 2009). However, several extremely resistant immunoreactive proteins still remained in some legumes, such as lentil and chickpea, even after these extreme treatments. Similar results have been found using instant controlled pressure drop (DIC®), a procedure that combines heat and steam pressure as in autoclaving (Cuadrado et al. 2011; Guillamón et al. 2008). In the present chapter, we have summarized the impact of DIC technology on the IgE-binding capacity of proteins from legumes such as lupine, peanut (raw and roasted), lentil, chickpea, and soybean. Its effect is compared with that produced by other thermal treatments, with and without pressure.

2 Major Legumes and Their Predominant Allergens

Legumes are a rich source of protein and other nutritious elements that have functional properties, which is why they are an important ingredient in manufactured foods today. Indeed legumes have become an important ingredient of the human diet throughout the world (Duranti 2006). Unfortunately, they are one of the common types of food that have the potential to elicit an allergenic response. Some common allergenic legumes are peanut, soybean, lentil, lupine, pea, chickpea, red gram, and black gram. Several legume allergens have been identified and characterized as belongings to the cupin superfamily of storage proteins (Mills et al. 2002). The major allergens of soybean and peanut have been extensively studied and Ara h 1, a 65 kDa glycoprotein belonging to the vicilin family, is one of the best characterized (Burks et al. 1998). Several legume allergens are strongly resistant to digestion and food processing, thus strengthening their allergenic potential (Mills et al. 2004). Fewer investigations have been carried out on lentil and chickpea allergens. No chickpea allergens have been identified and only three lentil allergens have been characterized until now. Research is being performed by various groups to identify more legume allergens (Verma et al. 2012).

The pattern of legume sensitization varies in different parts of the world, probably due to the genetic status of individuals, consumption habits, or maybe the involvement of other factors. A higher prevalence of peanut allergy is found in the UK, France, and North America, whereas a major incidence of soybean allergy has been reported in south-east Asia. Lentil, chickpea, and lupine allergic reactions are more widespread in the Mediterranean area (Crespo and Rodríguez 2003).

According to UE regulations it is mandatory to label a list of 14 groups of potential allergenic foods. These include peanut and soybean, which were among the first to be included, and they have been followed more recently by lupine and any ingredients derived from it (Commission Directive 2006/142/EC).

3 Effect of Thermal Processing on Legume Allergenic Proteins

Foods are subjected to thermal and nonthermal processing methods to improve their quality, preservation, safety, and suitability for specific product applications. The degree of processing affects digestibility, solubility, and other related parameters. During processing, proteins can form oligomers, become denatured, aggregate, fragment, and reassemble, and these changes most often reduce solubility (Maleki 2004). Processing can alter the overall IgE-binding profiles of a particular extract, which can become more or less antigenic or result in new allergens (neoallergens) (Schmitt et al. 2010). Therefore, the study of processing is necessary to assess the allergenicity of existing and newly introduced foods (Wal 2003). The effect of thermal processing mainly depends on temperature and duration. Moreover, alteration in the structure of a protein also depends on interactions with other food matrix constituents. In general, when the temperature is around 70–80 °C a loss of secondary structure occurs, whereas at 80–90 °C the formation of new bonds and rearrangements of disulfide bonds occur. At higher temperatures (90–100 °C) there is aggregate formation (Davis and Williams 1998).

Studies demonstrated that extracts from roasted peanut bind IgE at approximately 90-fold higher levels than those from raw peanuts (Maleki et al. 2000; Chung et al. 2003). These studies showed that the major allergens from roasted peanuts, Ara h 1 and Ara h 2, undergo structural alterations that enhance their allergenic properties. Heat treatment probably increases the digestibility of proteins, so their absorption through the gastrointestinal tract may also increase, decreasing the possibility for an allergenic protein to elicit an allergenic response. However, in some cases thermal processing may reduce the digestibility of a particular allergen or neoantigens may be formed that were not originally present. This general phenomenon may enhance the allergenic problem in sensitized patients and the neoantigens may also present an additional problem. One major factor responsible for the formation of neoantigens is the Maillard reaction, i.e., the interaction of protein components with sugar residues upon heating, generating sugar conjugated protein derivatives which enhance the allergenicity of proteins (Maleki et al. 2000).

IgE antibodies recognize and interact with epitopes present on allergenic proteins. IgE-binding epitopes can be either linear or conformational. In linear epitopes, the amino acids are arranged in linear order along the polypeptide chain, while in the conformational epitopes amino acids that are far apart in the primary sequence may come together during the folding of the polypeptide chain. Linear epitopes may be more problematic compared to the conformational ones, as the former are mostly resistant to heat treatment. Thermal processing mainly affects conformational epitopes as heat can break the bonds. Refolding allows the formation of native conformational epitopes but a few new allergens may be formed, requiring further efforts to minimize the risk associated with these neoantigens (Sathe and Sharma 2009). Thus, thermal processing as well as other processing events can dramatically alter the structure, function, and allergenicity of foods. Incomplete knowledge of the allergens in processed foods increases the complexity of food allergy diagnosis.

4 Effect of DIC on Immunoreactivity of Legume Proteins

In accordance with the aim of this chapter, the changes in the IgE-binding capacity of lupine, raw and roasted peanut, lentil, chickpea, and soybean proteins produced by DIC technology will be summarized.

All DIC treatments were carried out according to the experimental design developed by Haddad et al. (2001). Briefly, the moistened product is placed in a processing chamber and exposed to steam pressure (up to 8 bar) at high temperature (up to 170 °C) for a relatively short time (a few seconds to some minutes). An instant pressure drop towards a vacuum at about 50 mbar follows this high temperature–short time stage. This abrupt pressure drop simultaneously provokes an autovaporization of part of the water in the product and an instantaneous cooling, which stops thermal degradation. Whole seeds of lupine, raw and roasted peanut, lentil, chickpea, and soybean were treated at different pressures for different time periods, using a 12 or 22 central point composite design (4 or 10 repetitions, respectively). Some DIC-treated samples were selected for SDS-PAGE and immunoblotting studies: 3 and 6 bar for 1 and 3 min, with a constant initial water content of 50 g of water per 100 g of dry matter.

4.1 Effect of DIC on Lupine (L. albus) *Allergens*

White lupine is considered to be a rich source of protein with a high lysine content and is increasingly used both for its multifunctional properties and its potential hypocholesterolemic and hypoglycemic effects (Duranti 2006). However, lupine seed flour has been reported to be a causative agent of allergic reactions, especially in patients allergic to peanut (Moneret-Vautrin et al. 1999). Moreover, lupine seed proteins have proved to be an interesting model for the study of the thermal conformational stability of proteins under different pH conditions from both biochemical and technofunctional points of view (Duranti et al. 2000).

In a previous study, microwave cooking, boiling, and extrusion cooking produced minimal changes on IgE binding to lupine proteins; however, boiling in an autoclave at 2.6 bar (138 °C) for 20 min produced a significant decrease in allergenicity (Álvarez-Álvarez et al. 2005). Only two proteins of 23 and 29 kDa had IgE-binding capacity after autoclaving for 20 min, whereas autoclaving for 30 min completely abolished the IgE binding of these components, suggesting that lupine allergens are relatively heat stable. Two main allergens were subsequently identified as Lup 1 (conglutin β, 34.5 kDa, 7S protein) and Lup 2 (a basic subunit of conglutin α, 20 kDa, 11S protein) (Guillamón et al. 2010). Both were partially sequenced and their high degree of homology with major allergens from peanut, lentil, pea, and soybean could explain the IgE cross-reactivity of lupine with these legumes.

Guillamón et al. studied the effect of DIC treatment on lupine allergenicity (Guillamón et al. 2008). The SDS-PAGE and IgE-immunoblot analysis of raw and DIC-processed lupine proteins are shown in Fig. 1a, b. Raw lupine and lupine processed with DIC at 3 bar for 1 and 3 min had similar SDS and IgE-immunoblot band patterns; Lup 1 (34.5 kDa) and Lup 2 (20 kDa) were still present in these samples. A similar protein pattern was found by Álvarez-Álvarez et al. in autoclaved lupine seeds processed at 1.2 bar for 20 min (Álvarez-Álvarez et al. 2005). A major decrease in the number and intensity of the bands was observed when DIC was applied at 6 bar for 1 min, but Lup 1 (34.5 kDa) was still present. However, after 3 min at 6 bar neither Lup 1 nor Lup 2 were detected (Fig. 1b). This study confirmed the results of previous work that demonstrated the thermal resistance to autoclave treatment of the major lupine allergens (Álvarez-Álvarez et al. 2005). Both results suggest that lupine seeds treated with processing technologies combining heat and steam pressure, such as DIC and autoclaving, could almost completely eliminate the in vitro immunoreactivity of lupine.

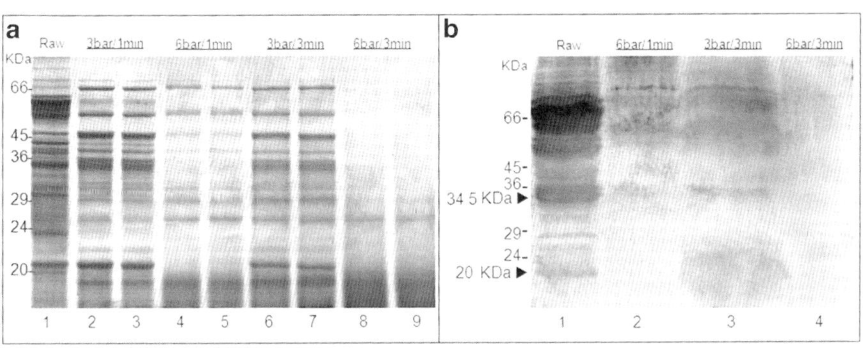

Fig. 1 (**a**) SDS-PAGE pattern of whole protein extract from raw (*lane 1*) and DIC-processed (*lanes 2–9*) lupine samples. (**b**) IgE immunoblot of whole protein extract from raw (*lane 1*) and DIC-processed (*lanes 2–4*) lupine samples, using a serum pool from patients sensitized to lupine (20 μg of protein per lane)

4.2 *Effect of DIC on Peanut* (A. hypogaea) *Allergens*

Peanut allergy is one of the most common IgE-mediated reactions to food because of its severity and lifelong persistence (Sicherer and Sampson 2007). Considerable effort has been spent in characterizing peanut allergens and 11 allergenic proteins have been identified until now (Ara h 1–Ara h 11). The major peanut allergens, Ara h 1 (65 kDa, vicilin) and Ara h 2 (17 kDa, conglutin), are recognized by 70–90 % of sensitized subjects (Burks et al. 1998) while Ara h 3 (11S legumin) is considered to play a lesser allergenic role (Rabjohn et al. 1999). Thermal treatment has a significant effect on peanut immunoreactivity. As previously mentioned, roasting peanut enhances its IgE-binding capacity (Maleki et al. 2000), while boiling decreases its allergenicity (Beyer et al. 2001).

According to Cabanillas et al., the IgE immunoreactivity of roasted peanut decreases significantly at extreme conditions of autoclaving (2.6 bar, 30 min) (Cabanillas et al. 2012). Results obtained by circular dichroism spectroscopy indicated that most of the α-helical structure was lost after autoclave treatment. The fact that many of the IgE-binding epitopes of major peanut allergens (Ara h 1, Ara h 2 and Ara h 3) are located on the α-helical regions may explain this decrease (Barre et al. 2007; Mueller et al. 2011).

When DIC treatment was used at 3 bar for 1 and 3 min and at 6 bar for 1 min with raw and roasted peanut proteins, it did not produce any relevant change in the immunoblot profile compared to untreated samples (Cuadrado et al. 2011). However, DIC treatment at 6 bar for 3 min resulted in a marked decrease in the protein bands of 65 kDa (putative Ara h 1) (Fig. 2) and no immunoreactive bands of less than 20 kDa were recognized. However, some bands (e.g., 37 kDa, Ara h 3 acid subunit) behave differently in raw and roasted peanuts (Fig. 2). Taking into account the reduction of in vitro immunoreactivity observed at the highest pressure and longest time, DIC seems to be more effective on the immunoreactivity of roasted peanut proteins than on that of raw peanuts. From the literature, it is apparent that changes in the immunoreactivity of peanut proteins following thermal treatment may be in large part due to a modification in the structure and reactivity of each individual peanut allergen and also to their interaction with the food matrix, although more studies are required to provide a fuller understanding of this question.

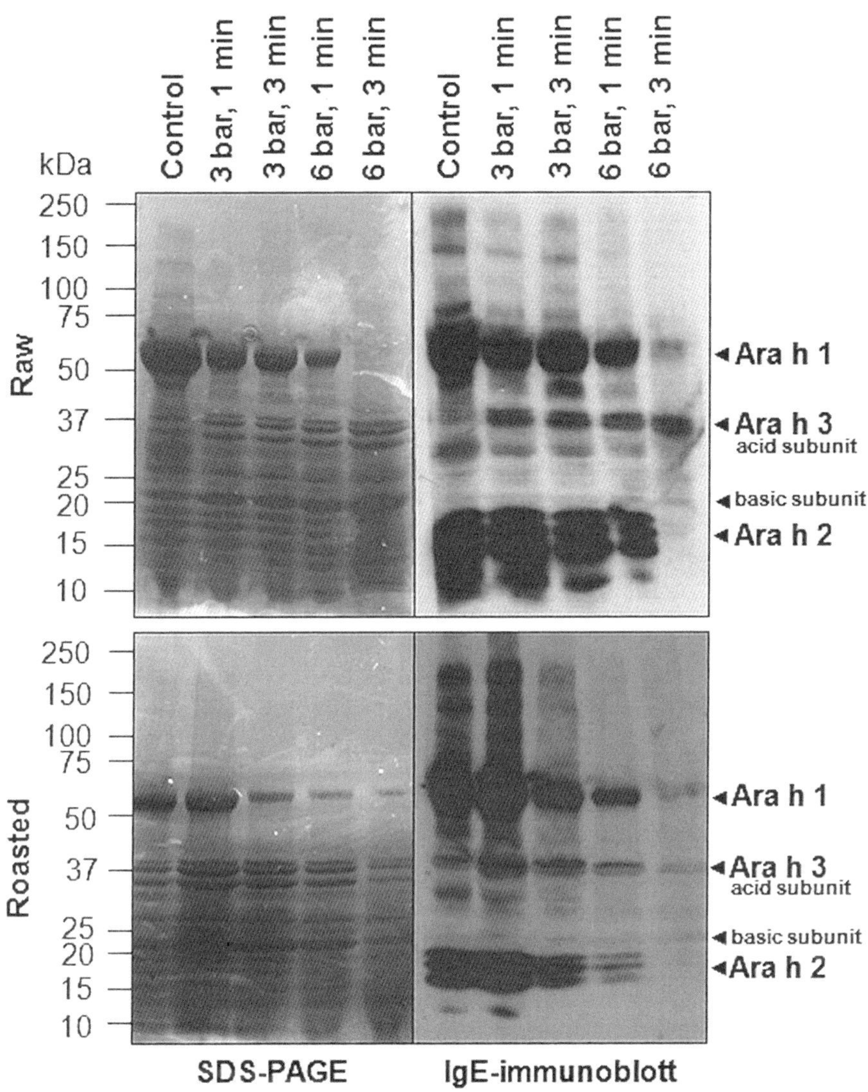

Fig. 2 SDS-PAGE patterns and IgE immunoblot from control and DIC-processed samples of raw and roasted peanuts. The pool serum used was from subjects sensitized to peanut (20 μg of protein per lane)

4.3 Effect of DIC on Lentil (L. culinaris) *Allergens*

Lentil is commonly consumed in Mediterranean areas and it has been reported as a cause of IgE-mediated hypersensitivity reactions, particularly in pediatric patients. Multiple IgE-binding allergens have been detected in both raw and boiled lentil but studies investigating the allergenicity of lentils with well-documented clinical sera are scarce. Two major lentil allergens, Len c 1 (48 kDa vicilin) (López-Torrejón et al. 2003) and Len c 2 (66 kDa) were isolated from boiled lentils and characterized (Sánchez-Monge et al. 2000). Recently, a third allergen, Len c 3, has also been characterized (Akkerdaas et al. 2012).

Boiling lentil proteins does not seem to be an effective way of reducing their allergenic potential. However, autoclave treatment of lentil (1.2 and 2.6 bar, up to 30 min) significantly decreased the activity of IgE-binding proteins. Autoclaving at the highest pressure (2.6 bar) for 30 min significantly reduced the overall IgE-binding capacity, although extremely resistant immunoreactive proteins still remained even after this harsh treatment (Cuadrado et al. 2009).

When we studied the effect of DIC treatment on lentil, we observed that the control sample was composed of numerous immunoreactive bands with molecular weights of between 101 and 140 kDa, including a protein of 48 kDa (putative major allergen Len c1) (Fig. 3a) (Cuadrado et al. 2011). DIC processing at 3 and 6 bar for 1 and 3 min produced a marked decrease in the overall immunoreactivity (data not shown) but extreme DIC conditions (6 bar, 3 min) reduced the number of IgE-binding proteins in a manner similar to that observed with autoclave treatment under harsh conditions. After this kind of DIC treatment only four heat-stable allergenic proteins of MW 29, 48, 57, and 68 kDa were still present (Fig. 3a). These results confirmed the thermostability of some lentil allergens, which are also detected after autoclaving.

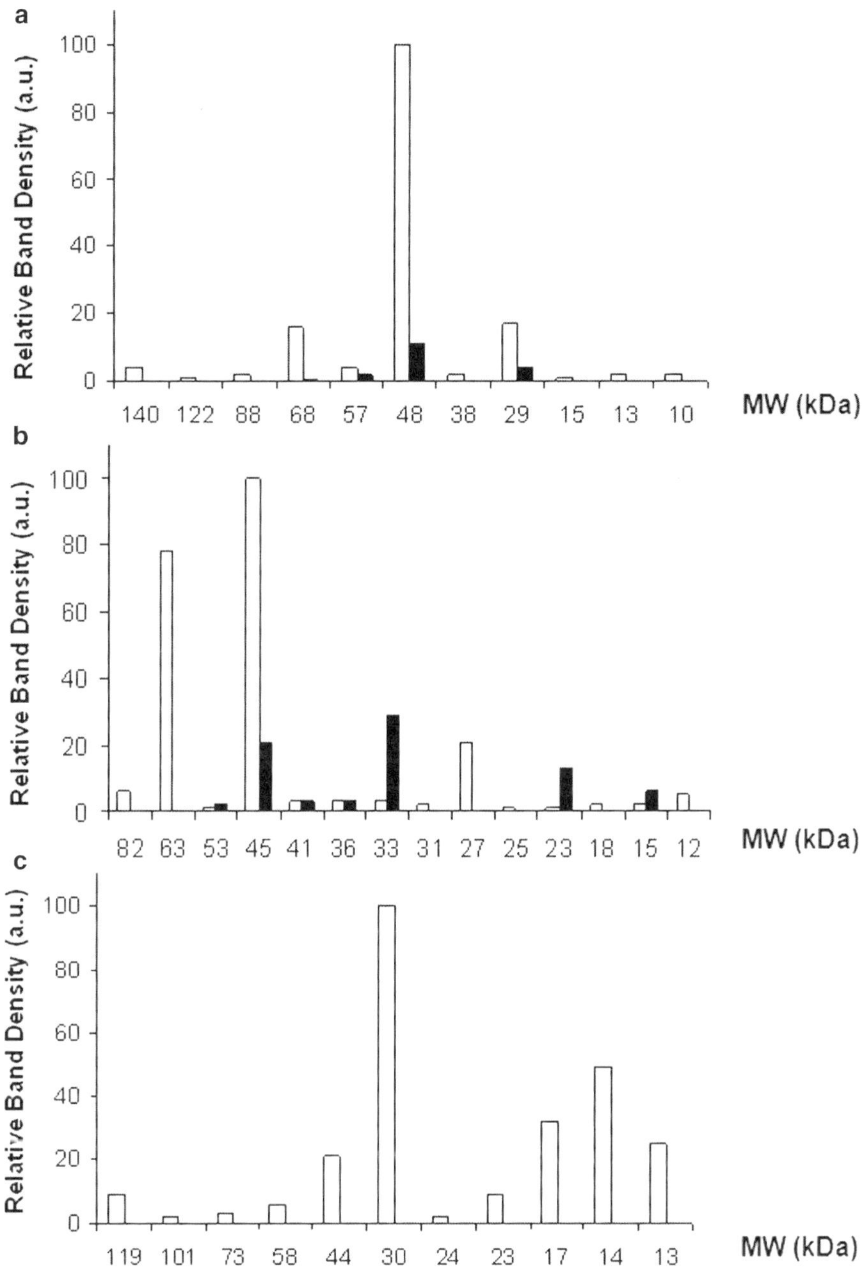

Fig. 3 Relative band density (a.u., arbitrary units) of immunoreactive proteins vs. band molecular weight (MW, kDa) obtained from IgE immunoblots from control and DIC-treated (6 bar, 3 min) samples of lentils (**a**), chickpeas (**b**), and soybeans (**c**). The pool serum used was from subjects sensitized to each legume. Where no *dark bands* are present the amount of material is below the limit of detection (control: white; DIC: black)

4.4 *Effect of DIC on Chickpea* (C. arietinum) *Allergens*

Chickpea is an important source of proteins in several parts of the world. In Asian countries it is widely consumed in many traditional dishes. The high consumption rate of this crop in the Mediterranean area has also resulted in allergic problems in sensitive individuals (Crespo and Rodríguez 2003). Chickpea and lentil are the most common cause of allergic reactions to legumes in Spanish children and there is a cross-reactivity between them (Crespo et al. 1995). Some subjects allergic to this legume on ingestion also report symptoms when they inhale vapors from cooking chickpeas (Niphadkar et al. 1997). Previous studies have detected multiple IgE-binding bands in chickpea extracts in the molecular weight range of 10–106 kDa, of which the majority were found to be heat stable (Patil et al. 2001). So far, chickpea allergens have not been immunologically characterized; only two allergenic polypeptides from chickpea (2S albumin and 11S globulin) have been identified (Vioque et al. 1999).

Cuadrado et al. found multiple IgE-binding proteins in chickpea boiled for 30 min (Cuadrado et al. 2009). A decrease in the number and intensity of the bands was observed after autoclaving (1.2 bar, 12 min) and the immunoreactivity decreased as pressure and time increased. At 2.6 bar (30 min) only two bands (19 and 16 kDa) were still detected.

The effect of DIC treatment on allergenic proteins from chickpea and lentil was similar. Untreated chickpea had numerous IgE-binding proteins with molecular weights of between 12 and 82 kDa (Fig. 3b). The immunoreactive band pattern after DIC treatment at 6 bar for 3 min also showed a marked decrease in the number and intensity of IgE-binding proteins. However, in the chickpea experiment, there were no apparent distinctions among the different pressure and time conditions used (data not shown), and more heat-stable immunoreactive proteins were still present at the same extreme DIC conditions (6 bar, 3 min) (Fig. 3b).

4.5 *Effect of DIC on Soybean* (G. max) *Allergens*

Soybean and peanut are the two main legumes involved in hypersensitive responses in numerous countries. Soybean, with around 21 known allergenic proteins, is widely consumed throughout the world. It is mainly used as an ingredient in formulated foods, meat/poultry products, together with bakery, pastry, and dairy products, and has many pharmaceutical and industrial uses (Endres 2001). The large amount of soybean consumed has also been associated with a high risk of allergy for consumers and the prevalence rate in the general population is around 0.3–0.4 %. Several major allergens have been identified in soybean, namely, P34 (Gly m Bd 30 K), Gly m 1, Gly m 2, Gly m 3, Gly m 4, and Gly m Bd 28 K (Verma et al. 2012). Given the commercial use of soybean protein in food products, which frequently include thermally processed proteins, it is of great importance to investigate the effect of heat treatment on the main allergens of this legume. Burks et al. did not find any relevant decrease in IgE binding after heating soy proteins at various temperatures and for various times (Burks et al. 1991). Wilson et al. (2005) concluded that several procedures are needed to eliminate soybean allergenicity, particularly that of P 34, the major allergenic protein.

To evaluate the effect of DIC, untreated soybean was used as a control (Cuadrado et al. 2011). All the major allergens and other minor immunoreactive proteins (13–119 kDa) were detected in the control soybean (Fig. 3c). Although DIC treatment at 3 bar for 1 and 3 min resulted in a slight reduction in the soybean immunoreactive bands (data not shown), when the pressure was increased to 6 bar and applied for 3 min, no immunoreactive proteins could be detected on the immunoblot pattern (Cuadrado et al. 2011) (Fig. 3c). Similar results were found when this legume was autoclaved at 2.6 bar for 30 min (Cuadrado et al. 2007). The DIC technique employed here had the strongest effect on the immunoreactive proteins of soybean compared to the other legumes studied. The short processing time (3 min) represents an advantage for future potential applications in the food industry. Elimination of allergenic proteins via processing could eventually enhance the safety of soybean products, making them available for soy-sensitized individuals.

5 Conclusions

According to Thomas et al. (2007), food processing may impact the potential allergenicity of proteins, although there are no general rules regarding how allergenic foods respond to physical, chemical, or biochemical processing methods. The modifications may result in a loss of organized structure and protein denaturation. Moreover, it has been demonstrated that the degree of processing can dramatically affect digestibility, solubility, and other parameters related to IgE reactivity. In some allergenic proteins, the epitopes are destroyed but they are unaltered in others. Thus, processing can alter the overall IgE-binding profiles of legume proteins.

This chapter summarized the effect of DIC treatment (steam pressure, high temperature, and short time) on different legumes and compared it with other thermal procedures. DIC treatment significantly decreased the overall immunoreactivity of lentil, chickpea, and peanut, mainly roasted, and almost completely eliminated IgE recognition of lupine and soybean proteins. These DIC treated legumes could constitute an alternative to intact proteins in the development of different food products. However, as in vitro IgE reactivity is an indicator of potential in vivo allergenicity, further in vitro assays and in vivo clinical data are required to confirm that DIC treatment can reduce the in vivo allergenicity of these legumes. Only after such studies could these putative hypoallergenic foods be safely consumed and even utilized as a desensitizing food.

References

Akkerdaas J, Finkina E, Balandin S, Santos Magadan S, Knulst A, Fernández-Rivas M, Asero R, van Ree R, Ovchinnikova T (2012) Lentil (Lens culinaris) lipid transfer protein Len c 3: a novel legume allergen. Int Arch Allergy Immunol 157(1):51–57. doi:10.1159/000324946

Álvarez-Álvarez J, Guillamón E, Crespo JF, Cuadrado C, Burbano C, Rodríguez J, Fernández C, Muzquiz M (2005) Effects of extrusion, boiling, autoclaving, and microwave heating on lupine allergenicity. J Agric Food Chem 53(4):1294–1298. doi:10.1021/jf0490145

Barre A, Jacquet G, Sordet C, Culerrier R, Rouge P (2007) Homology modelling and conformational analysis of IgE-binding epitopes of Ara h 3 and other legumin allergens with a cupin fold from tree nuts. Mol Immunol 44(12):3243–3255. doi:10.1016/j.molimm.2007.01.023

Besler M, Steinhart H, Paschke A (2001) Stability of food allergens and allergenicity of processed foods. J Chromatogr B Biomed Sci Appl 756:207–228

Beyer KB, Morrow E, X-M L, Bardina L (2001) Effects of cooking methods on peanut allergenicity. J Allergy Clin Immunol 107:1077–1081

Burks A, Williams L, Helm RM, Thresher W, Brooks J, Sampson H (1991) Identification of soy protein allergens in patients with atopic dermatitis and positive soy challenges: determination of change in allergenicity after heating or enzyme digestion. Adv Exp Med Biol 289:295–307

Burks A, Sampson H, Bannon G (1998) Peanut allergens. Allergy 53:725–730

Cabanillas B, Maleki SJ, Rodríguez J, Burbano C, Muzquiz M, Jimenez M, Pedrosa MM, Cuadrado C, Crespo JF (2012) Heat and pressure treatments effects on peanut allergenicity. Food Chem 132(1):360–366. doi:10.1016/j.foodchem.2011.10.093

Chung S-Y, Reed S (2011) Removing peanut allergens by tannic acid. Food Chem 134:1468–1473

Chung S, Butts C, Maleki SJ, Champagne E (2003) Linking peanut allergenicity to the processes of maturation, curing, and roasting. J Agric Food Chem 51:4273–4277

Commision Directive 2006/142/EC of the European Parlament and of the Council (2006) Official Journal of the European Union L 368/110.

Crespo JF, Rodriguez J (2003) Food allergy in adulthood. Allergy 58:98–113

Crespo JF, Pascual C, Burks A, Helm R, Martín-Esteban M (1995) Frequency of food allergens in a paediatric population from Spain. Pediatr Allergy Immunol 6:39–43

Cuadrado C, Cabanillas B, Pedrosa MM, Varela A, Guillamón E, Muzquiz M, Crespo JF, Rodríguez J (2007) Effect of heat treatments on legume allergenicity. . Paper presented at the Integrating Legume Biology for Sustainable Agriculture. Proceedings of the 6th European Conference on Grain Legumes, Lisbon, Portugal, 12–16 November

Cuadrado C, Cabanillas B, Pedrosa MM, Varela A, Guillamón E, Muzquiz M, Crespo JF, Rodríguez J, Burbano C (2009) Influence of thermal processing on IgE reactivity to lentil and chickpea proteins. Mol Nutr Food Res 53(11):1462–1468. doi:10.1002/mnfr.200800485

Cuadrado C, Cabanillas B, Pedrosa MM, Muzquiz M, Haddad J, Allaf K, Rodríguez J, Crespo JF, Burbano C (2011) Effect of instant controlled pressure drop on IgE antibody reactivity to peanut, lentil, chickpea and soybean proteins. Int Arch Allergy Immunol 156(4):397–404. doi:10.1159/000324443

Davis P, Williams S (1998) Protein modification by thermal processing. Allergy 53:102–105

Duranti M (2006) Grain legume proteins and nutraceutical properties. Fitoterapia 77:67–82

Duranti M, Sessa F, Scarafoni A, Bellini T, Dallocchio F (2000) Thermal stabilities of lupine seed conglutin ç protomers and tetramers. J Agric Food Chem 48:1118–1123

Endres J (ed) (2001) Soy protein products-characteristics, nutritional aspects, and utilization. AOCS, Champaign

Guillamón E, Burbano C, Cuadrado C, Muzquiz M, Pedrosa MM, Sánchez M, Cabanillas B, Crespo JF, Rodríguez J, Haddad J, Allaf K (2008) Effect of an instantaneous controlled pressure drop on in vitro allergenicity to lupines (Lupinus albus var Multolupa). Int Arch Allergy Immunol 145(1):9–14. doi:10.1159/000107461

Guillamón E, Rodríguez J, Burbano C, Muzquiz M, Pedrosa MM, Cabanillas B, Crespo JF, Sancho A, Mills C, Cuadrado C (2010) Characterization of lupine major allergens (*Lupinus albus* L). Mol Nutr Food Res 54:1168–1676

Haddad J, Louka N, Gadouleau M, Juhel F, Allaf K (2001) Application du nouveau procédé de séchage/texturation par Détente Instantanée Contrôlée DIC aux poissons: impact sur les caractéristiques physico-chemiques du produit fini. Sci Aliment 21:481–498

Leterme P (2002) Recommendations by health organizations for pulse consumption. Br J Nutr 88:239–242

López-Torrejón G, Salcedo G, Martín-Esteban M, Diaz-Perales A, Pascual CY, Sánchez-Monge R (2003) Len c 1, a major allergen and vicilin from lentil seeds: protein isolation and cDNA cloning. J Allergy Clin Immunol 112(6):1208–1215. doi:10.1016/j.jaci.2003.08.035

Maleki SJ (2004) Food processing: effects on allergenicity. Curr Opin Allergy Clin Immunol 4 (3):241–245. doi:10.1097/00130832-200406000-00018

Maleki SJ, Chung SY, Champagne ET, Raufman JP (2000) The effects of roasting on the allergenic properties of peanut proteins. J Allergy Clin Immunol 106(4):763–768. doi:10. 1067/mai.2000.109620

Mills C, Jenkins J, Marigheto N, Belton P, Gunning A, Morris V (2002) Allergens of the cupin superfamily. Biochem Soc Trans 30(6):925–929

Mills C, Jenkins JA, Alcocer M, Shewry P (2004) Structural, biological, and evolutionary relationships of plant food allergens sensitizing via the gastrointestinal tract. Crit Rev Food Nutr 44:379–407. doi:10.1080/10408690490489224

Mills C, Mackie A, Burney P, Beyer K, Frewer L, Madsen C, Botjes E, Crevel R, van Ree R (2007) The prevalence, cost and basis of food allergy across Europe. Allergy 62(7):717–722

Mondoulet L, Paty E, Drumare M-F, Sea A-L (2005) Influence of thermal processing on the allergenicity of peanut proteins. J Agric Food Chem 53:4547–4553

Moneret-Vautrin D, Guérin L, Kanny G, Flabbee J, Frémont S, Morisset M (1999) Cross-allergenicity of peanut and lupine: the risk of lupine allergy in patients allergic to peanuts. J Allergy Clin Immunol 04(4 Pt 1):883–888

Mueller GA, Gosavi RA, Pomes A, Wuenschmann S, Moon AF, London RE, Pedersen LC (2011) Ara h 2: crystal structure and IgE binding distinguish two subpopulations of peanut allergic patients by epitope diversity. Allergy 66(7):878–885. doi:10.1111/j.1398-9995.2010.02532.x

Niphadkar PV, Patil SP, Bapat MM (1997) Chickpea induced anaphylaxis. Allergy 52:480–482

Patil SP, Niphadkar PV, Bapat MM (2001) Chickpea: a major food allergen in the Indian subcontinent and its clinical and immunochemical correlation. Ann Allergy Asthma Immunol 87(2):140–145

Rabjohn P, Helm E, Stanley J, West C, Sampson H, Burks A, Bannon G (1999) Molecular cloning and epitope analysis of the peanut allergen Ara h 3. J Clin Invest 103:535–542

Sánchez-Monge R, Pascual CY, Díaz-Perales A, Fernández-Crespo J, Martín-Esteban M, Salcedo G (2000) Isolation and characterization of relevant allergens from boiled lentils. J Allergy Clin Immunol 106(5):955–961. doi:10.1067/mai.2000.109912

Sathe SK, Sharma GM (2009) Effects of food processing on food allergens. Mol Nutr Food Res 53 (8):970–978. doi:10.1002/mnfr.200800194

Schmitt D, Nesbit J, Hurlburt B, Cheng H, Maleki SJ (2010) Processing can alter the properties of peanut extract preparations. J Agric Food Chem 58(2):1138–1143. doi:10.1021/jf902694j

Sicherer S, Sampson H (2007) Peanut allergy: emerging concepts and approaches for an apparent epidemic. J Allergy Clin Immunol 120(3):491–503

Thomas K, Herouet-Guicheney C, Ladics G, Bannon G, Cockburn A, Crevel R, Fitzpatrick J, Mills C, Privalle L, Vieths S (2007) Evaluating the effect of food processing on the potential human allergenicity of novel proteins: international workshop report. Food Chem Toxicol 45:1116–1122

Verma A, Kumar S, Das M, Dwivedi P (2012) A comprehensive review of legume. Clin Rev Allerg Immunol. doi:10.1007/s12016-012-8310-6

Vioque J, Sánchez-Vioque R, Clemente A, Pedroche J, Bautista J, Millán F (1999) Purification and partial characterization of chickpea 2S albumin. J Agric Food Chem 47(4):1405–1409. doi:10.1021/jf980819k

Wal J (2003) Thermal processing and allergenicity of foods. Allergy 58:727–729

Wilson S, Blaschek K, Gonzalez de Mejia E (2005) Allergenic proteins in soybean: processing and reduction of P34 allergenicity. Nutr Rev 63:47–58

DIC Decontamination of Solid and Powder Foodstuffs

Sabah Mounir, Nsren Albitar, and Karim Allaf

1 Introduction

Stricter safety standards are becoming an increasingly restrictive element in food production. Furthermore, the precautionary procedures adopted in various manufacturing industries to produce healthy foods are very important but in no way sufficient to ensure complete safety from different microbiological contaminations. Indeed, cleaning processes are never perfect and there is an acute need for a specific microbiological decontamination step to be included in the industrial process, even in the most developed countries.

Although many new techniques have been studied, optimized, and sometimes used to protect food from microbiological deterioration, ultrahigh temperature (UHT) is usually considered to be the most valuable decontamination process. Other techniques such as microwaves and ohmic heating sometimes help to improve the performance of this type of operation. Physical treatments, such as pulsed electromagnetic field (PEF), ultraviolet (UV) treatment, and ultrahigh pressure, have been studied, optimized, and applied. Although the use of the broadest possible range of technologies has been successfully implemented with several new types of treatment, their use remains very limited. Chemical additives and ionization are increasingly being rejected by consumers and limited by international standards.

S. Mounir
Department of Food Science, Faculty of Agriculture, Zagazig University, Zagazig, Egypt

N. Albitar
Faculty of Chemical and Petroleum Engineering, El-Baath University, Homs, Syria

K. Allaf (✉)
Laboratory of Engineering Science for Environment (LaSIE FRE 3474) CNRS,
17042 La Rochelle Cedex 01, University of La Rochelle, La Rochelle, France
e-mail: kallaf@univ-lr.fr

T. Allaf and K. Allaf (eds.), *Instant Controlled Pressure Drop (D.I.C.) in Food Processing: From Fundamental to Industrial Applications*, Food Engineering Series,
DOI 10.1007/978-1-4614-8669-5_5,

There are multiple barriers to achieving good decontamination in terms of technical performance (efficiency of microorganism destruction, energy consumption) and an efficient preservation of the attributes and quality of the end product (suitable safeguarding of biochemical, nutritional, and sensory contents). Heat treatment remains the major industrial operation used to destroy microorganisms, and UHT has been a real success in the food industry because it can clearly satisfy these two contradictory constraints. With specific optimization needed for each case, conventional UHT is not limited to pasteurization and also includes sterilization. However, UHT is only used with liquids as materials that can pass through a pump using exchangers for both the heating and/or cooling stages.

A large range of dried foodstuffs such as mushrooms, spices, herbs, onions, garlic, and flour are known to have a high microbial load, sometimes combined with an infestation with insects. This is often the result of traditional methods of harvesting, drying, grinding, storage, etc. Very few relevant decontamination technologies are appropriate for this type of product. Heat destruction of microbiological organisms in solids is faced with several types of difficulty. Conventional exchangers are not appropriate for the kind of heat transfer required. Conventional steam treatments often need long periods of rising and falling temperature. During the heating and cooling stages, a strong temperature gradient is produced and the products may suffer from a lack of homogeneity, which typically involves damage to the end product and reduces its overall quality. Although microwave heating is easy, it is almost impossible to obtain a uniform product and this method is not appropriate for dried food products, whatever their shape and size.

The use of both chemical and room-temperature specific gases such as ethylene oxide or propylene oxide is a long-standing practice. Nevertheless, international standards (European, US, Japanese) are increasingly stringent, mainly with regard to residual chemical molecules generated or introduced into the treated product, which in general are completely banned; ethylene oxide, for example, has been prohibited in France since 1990. Consumer attitudes are significantly reinforcing this tendency.

Nuclear gamma irradiation has been considered for a very long time to be the best and most convenient decontamination method, mainly for solid and/or powder foodstuffs. This treatment has proved to be highly effective and very convenient since the product can be treated in its airtight packaging, avoiding any recontamination. The dose for food depends on the desired effect: "low" dose irradiation (50–150 Gy) does not decontaminate and is used only to inhibit sprouting of potatoes and onions. Food sterilization (e.g., precooked meals) requires much higher doses (10–50 kGy). Nevertheless different authors have noted some negative effects of such treatment. Because of the radiation energy, ionization can remove electrons from atoms and break molecular bonds, leading to the formation of highly reactive free radicals. New molecules could thus appear in the food as a result of chemical recombination. The irradiation of lipids causes the formation of cyclobutanones whose toxicity is well known. Such chemical toxicity has been shown to have very disturbing effects (including chromosomal damage).

This type of irradiation treatment eliminates microorganisms in food, including those with useful features. However, some acids and vitamins (including A, B1, B6, B12, C, E, K, PP, and folic acid) can be damaged by radiation treatment, depending on the dose and the radiosensitivity of the molecules. High-dose irradiation destroys microorganisms but does not remove the toxins previously produced, which are often responsible for many foodborne illnesses. Moreover, some authors have highlighted the risk of a particular mutation which can be induced by irradiation in insects and bacteria. In addition to these scientific and technical aspects and despite a substantial marketing effort deployed by the industry to attest to the safety of irradiation (proved only for doses up to 10 kGy), ionizing radiation suffers from a very bad image due to the confusion between radiation and radioactivity. Rejection by consumers has been particularly reinforced since the introduction in 2001 of mandatory labeling of all products processed by ionizing radiation. These elements and the relatively high cost of the equipment explain its very low level of approval (Allaf et al. 2011).

2 DIC as an UHT Decontamination

"Instant controlled pressure drop" (DIC) is typically presented as a specific UHT treatment for solid and powder foods. Generally, heat treatment by DIC is a short controlled process carried out at high temperature (Fig. 1).

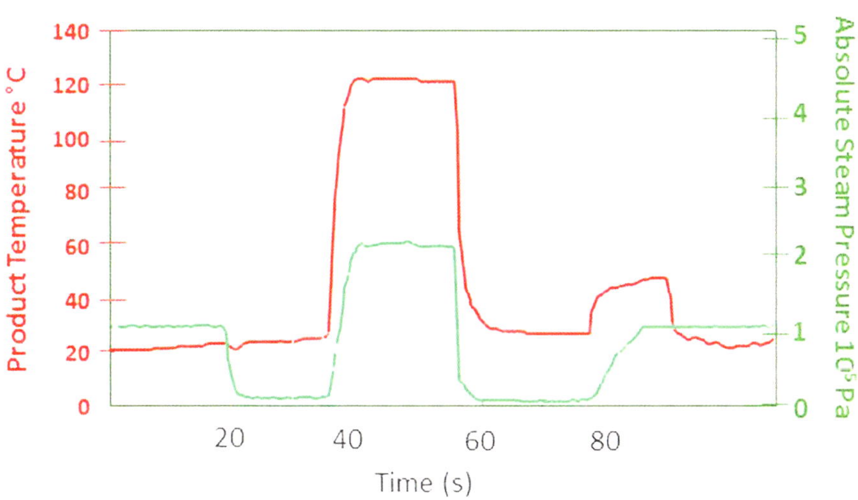

Fig. 1 A DIC treatment cycle: change in pressure and temperature versus time and various steps

The initial heating of the product can be substantially intensified and the time during which the temperature is rising can be radically reduced, since an initial vacuum step is inserted just before injecting high temperature–high pressure steam. The close contact between the steam and the surface of the product is indispensable to bring the product temperature to the same level as the steam temperature within a few seconds. The need to heat very rapidly can then "easily" be achieved. Cooling is obtained through the use of an instant controlled pressure drop towards a vacuum and can be measured on a 20/200 ms time frame. With this type of cooling it is possible to lower the temperature by decreasing the vacuum pressure. Typically, for DIC the vacuum is at about 5 kPa and the equilibrium temperature about 35 °C. Recent work has demonstrated the possibility of significantly reducing the temperature of the product to below the equilibrium level. This treatment has been studied in depth and the authors showed that it had a similar thermal effect to UHT for liquids. The treatment temperature is usually between 100 and 150 °C, while treatment time can be limited to 5 s. Furthermore, two other thermal stress impacts (short-time heating and instant cooling) seem to be very effective in the destruction of microorganisms. The instant autovaporization also induces micromechanical constraints that act on the microorganism cell walls and more specifically on the spore wall.

Indeed, specific research on the impact of instant pressure drop on cell structures and other biological structures is both of scientific interest and has an obvious technological importance. Products are initially placed under a vacuum and then saturated or superheated steam is injected at a fixed pressure which rises to an absolute pressure of 7 or 8 bar. This results in very rapid heating, mainly through condensation on the inner (fractal) surface of the product. With 5 mm thick dry products this can be achieved in less than 3 s. The heating time is completely controlled by inserting an intermediate vacuum stage at around 5 kPa of absolute pressure.

After the heating stage, which is usually carried out between 100 and 150 °C for 5–60 s, a second vacuum stage is immediately performed and lasts less than 100 ms, i.e., a decompression rate higher than 5 MPa/s. Thanks to autovaporization, the temperature of the product drops abruptly to very low levels, which must be lower than the equilibrium temperature of the water, which in our case is 33 °C.

Other versions of this process consist in achieving proper heating using high temperature adiabatic airflow (or inert gas). Moreover, heating can be achieved through contact with hot plates or even by microwaves. Coupled with vibration or mixing, the products attain the treatment temperature in a more uniform way. The application of an instant pressure drop towards a vacuum generates instant cooling by autovaporization, thus partly removing the water in the product.

This treatment perfectly reflects a UHT-type heat treatment but is here applied to solids instead of liquids. Thus, studies conducted by Debs-Louka et al. (1999) on the DIC-mediated microbial destruction of *Bacillus stearothermophilus* spores gave a value of $D_{121.1\ ^\circ C}$=2.6 min (Fig. 2) and a thermoresistance z of 7.6 °C, instead of 4.2 min and 8.8 °C, respectively (Le Jean et al. 1994).

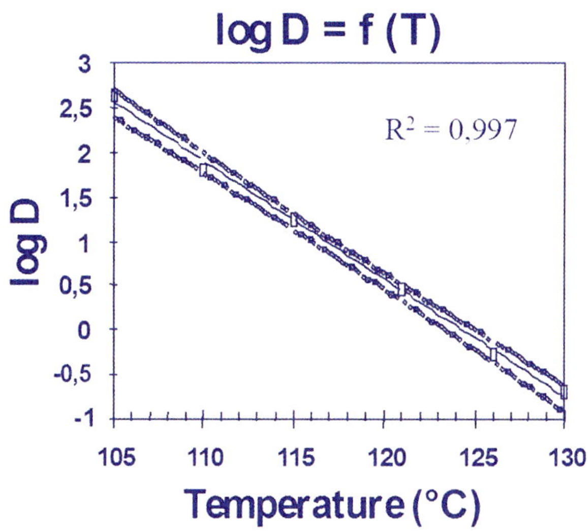

Fig. 2 Change in log *D* versus temperature for *Bacillus stearothermophilus* spores

3 DIC Thermomechanical Impacts

It was also possible to prove that the effect of the instant pressure drop in DIC was not confined to the impact of abrupt cooling. A thermomechanical effect that may lead to the explosion of microorganism cells (spores or vegetative forms) also occurs (Fig. 3).

Indeed, the higher the amount of "steam" generated by autovaporization within the cell and the shorter the pressure drop time, the more efficient the mechanical effect is.

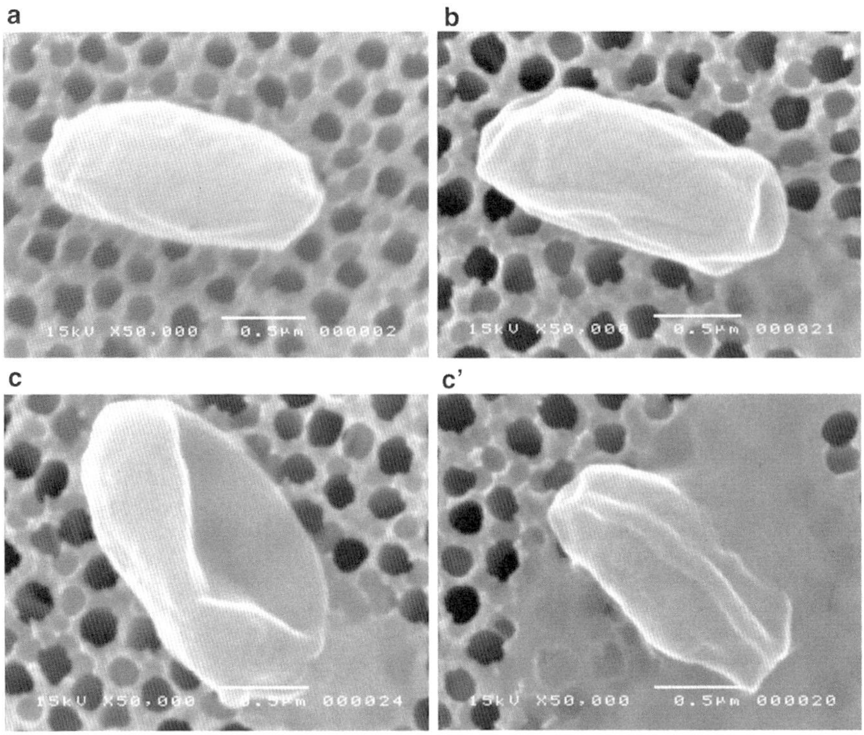

Fig. 3 Structure of *Bacillus stearothermophilus* spores: (*A*) Natural; (*B*) Heat treated at 130 °C for 30 s without a pressure drop; (*C and C'*) DIC treated at 130 °C for 30 s with an instantaneous pressure drop

4 Gas Mechanical Effect

Besides water, other molecules can be considered. The choice of molecule is closely related to its potential mechanical action, which is correlated to the difference in magnitude between the internal and the external pressure generated when the pressure instantaneously drops towards a vacuum; this normally depends on the rate of the pressure drop.

The choice of carbon dioxide was dictated by the relatively high dissolution capacity of this gas in microorganism cells during the high-pressure stage. For each fluid used the most important point was its capacity to generate a force capable of cracking and even breaking the cell walls. The treatment conditions were defined as being outside the supercritical range in order to avoid extraction consequences (Fig. 4). Other experimental approaches were carried out to identify the effect of the product's moisture content on the destruction of microorganisms by instantaneously dropping the pressure of CO_2.

Thus, experiments were conducted with *E. coli* and *S. cerevisiae* to investigate the influence of moisture content. Lin (1984) found that moisture content was vital to the antimicrobial action of CO_2. However, treatment carried out with carbon dioxide at 4 MPa for 195 min followed by an instant controlled pressure drop, with a moisture content of between 37 and 75 % dry basis, gave the same rate of destruction for the two microbial strains. Nevertheless, moisture content of about 6 % db was too low for inactivation.

Moisture content (% db dry basis)	CO_2 conditions (pressure–time)	log (N/N_0) (*Escherichia coli*)	log (N/N_0) (*Saccharomyces cerevisiae*)
37 % db—75 % db	5 MPa—300 min	−4.8	−4.3
6 % db	5 MPa—300 min	−0.09	−0.57

In the near future, further studies on the impact of high temperature will be carried out to investigate different aspects of high pressure/instant pressure drop with carbon dioxide.

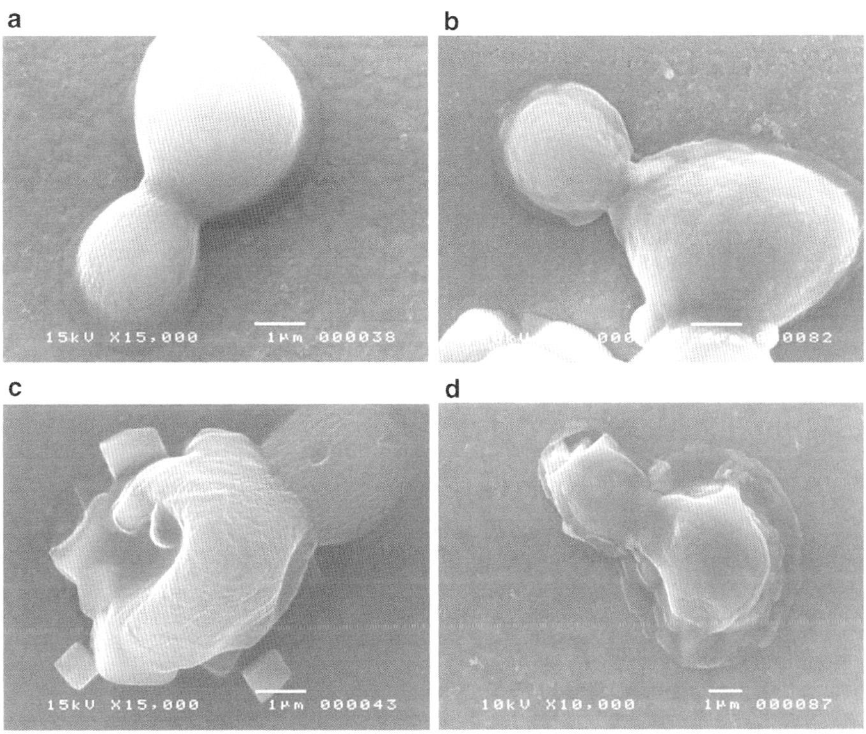

Fig. 4 *Saccharomyces cerevisiae* cell structure: (*A*) Natural; (*B*) Heat treated at 130 °C for 30 s without a pressure drop; (*C*) DIC treated at 105 °C for 30 s with an instantaneous pressure drop towards a vacuum (5 kPa; $\Delta P/\Delta t$ > 5 MPa/s); (*B'*) High-pressure carbon dioxide treatment at 55 bar for 5 h, without a pressure drop; (*C'*) High-pressure carbon dioxide treatment at 55 bar for 5 h, and ambient temperature, with a pressure drop towards a vacuum (5 kPa; $\Delta P/\Delta t$ > 5 MPa/s)

5 DIC Decontamination: Industrial Applications

Industrial applications have mainly consisted in adopting thermomechanical steam destruction of microorganisms while maintaining different parameters of food quality (texture, flavor, vitamins, protein activities, etc.). Research carried out by Allaf et al. (1998) and Debs-Louka et al. (1999) identified the different effects of this treatment and the well-established "temperature–time" couple for the level of decontamination. By quantifying the effect on various quality parameters, it was possible to obtain a very relevant and multidimensional optimization.

The major part of the industrial work using DIC treatment was subsequently carried out by the ABCAR-DIC process company. A wide-ranging variety of solids and powders, such as mushrooms, fruits, vegetables, meat and seafood, algae and microalgae, spices, and ginger, were effectively treated. The effect of decontamination by DIC has therefore been optimized at the industrial scale, according to various constraints related to the product and its requirements in terms of quality, the microorganisms that must be eliminated, and those that should be preserved.

6 Multi-cycle DIC Decontamination

The impact of the number of DIC cycles (various pressure drops towards a vacuum for the same total processing time) was quantified for dairy powder decontamination (Fig. 5; Table 1). The specificity of DIC treatment is related to the ability to define the degree of decontamination across the triumvirate "temperature/treatment time/number of pressure drops" instead of the conventional torque "temperature–treatment time." The impact of this specificity is immediate in terms of the quality of the finished product.

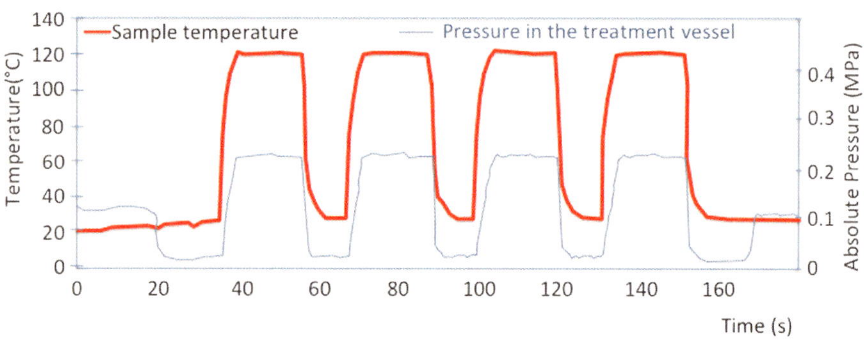

Fig. 5 Multi-cycle DIC decontamination treatment

Table 1 Multi-cycle DIC decontamination treatment conditions using hot air at 0.3 MPa, a vacuum of 5 kPa, and a pressure drop of about 1.5 MPa/s

Hot air	Temperature (°C)	Time (s)	Number of cycles	Time per cycle (s)
B1	135	360	6	60
B2	150	300	6	50
B3	135	420	6	70
B4	135	300	6	50
B5	144	360	8	45
B6	135	300	10	30
B7	135	300	6	50
B8	144	240	8	30
B9	144	300	4	75
B10	135	300	6	50
B11	144	240	4	60
B12	126	360	8	45

Cereus spores were inoculated at 10^4 CFU/g and analyses were performed in terms of both spores and vegetative forms, before and after DIC treatment. Statistical analysis (Statgraphics Plus) was mainly carried out on the decontamination ratio, which is the percentage of germs eliminated:

$$\text{Decontamination ratio (DR)} : \text{DR} = \frac{N_o - N}{N_o} \quad (1)$$

As the experiments were undertaken with an instant release of pressure, from high temperature towards the equilibrium temperature of water/vapor at 5 kPa, the operation was carried out in the "explosion conditions" defined by Lin et al. (1991, 1992). In the cases we studied, only 20–200 ms were necessary to reach the complete vacuum stage from 0.3 MPa ($\Delta P/\Delta t$ from 1.5 to 15 MPa/s) within the processing vessel. When the same treatment conditions were used but with a decompression time of 12 s ($\Delta P/\Delta t = 25$ kPa/s), no explosion effect occurred. Statistical analysis carried out on the results concerning the inhibition of vegetative and spore forms indicated that with multi-cycle DIC, the effects of processing temperature and total thermal processing time (t) were the most relevant parameters. However, the number of cycles (C) had a non-negligible effect. It is worth noting that the higher the T, t, and C, the greater the direct impact of multi-cycle DIC decontamination (Figs. 6 and 7).

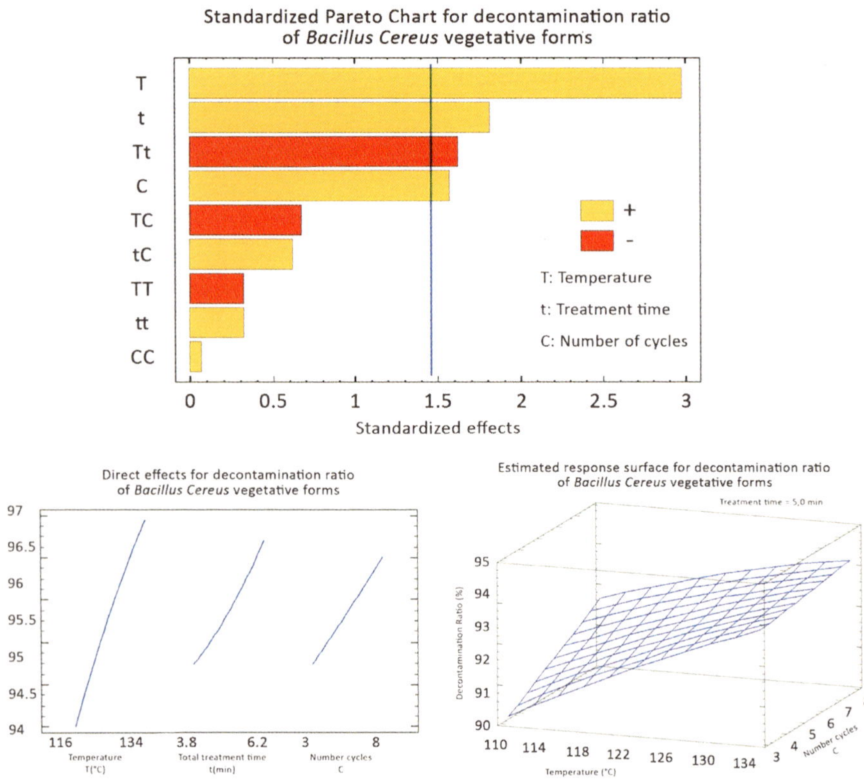

Fig. 6 Pareto chart, trends of main effects, and response surfaces of pressing temperature T (°C), total thermal treatment time t (s), and number of cycles (C) as multi-cycle DIC operating parameters from a five-level central composite rotatable RSM experimental design with ASR vegetative forms

It was then possible to establish empirical models of the MC-DIC decontamination ratio of both ASR vegetative and spore forms versus the DIC processing parameters, with $R^2 = 60.2\ \%$ and 76.77 %, respectively:

$$\mathrm{DR}_{\text{vegetative-ASR}} = -806 + 8T + 73t + 14C - 0.012T^2 - 0,64Tt - 0.13TC + 0.67t^2 + 0.91tC + 0.03C^2$$

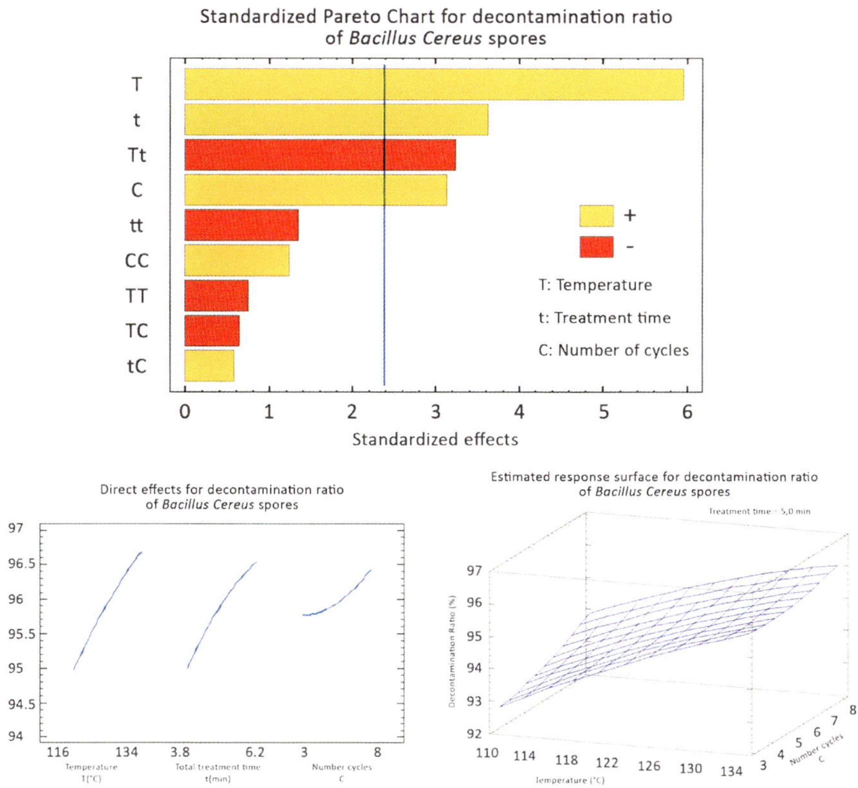

Fig. 7 Pareto chart, trends of main effects, and response surfaces of pressing temperature T (°C), total thermal treatment time t (s), and number of cycles (C) as multi-cycle DIC operating parameters from a five-level central composite rotatable RSM experimental design with ASR spores

$$DR_{spores\text{-}ASR} = 1,365 + 13T + 163t + 6C - 0.02T^2 - 1,1Tt - 0.1TC - 1.97t^2 + 0.66tC + 0.45C^2$$

These two models were used to optimize the multi-cycle DIC treatment parameters in order to reach the highest decontamination ratio while taking into account the preservation of quality. The higher the operating parameters T, t, and C, the higher the decontamination ratio. However, it is worth noting that the effect of C is specific because its positive impact on decontamination does not imply any thermal degradation of quality.

7 Conclusion

Instant controlled pressure drop (DIC) technology can be defined as a highly appropriate UHT-type decontamination process that can be applied to powders and other dry solid biological materials. This operation also involves thermal stress by both heating and, to a much greater extent, instant cooling. Another stress is induced by vapor constraints acting on the cell walls due to the rapid drop in pressure. This usually induces an explosion of the cell walls of both the vegetative and the spore forms.

Because of these different thermal and mechanical stresses, industrial uses of DIC are very effective for a wide variety of very sensitive biological products. Decontamination is usually optimized according to the product quality to be obtained. Since the number of cycles is a relevant operating parameter, competing with temperature and thermal processing time, multi-cycle DIC treatment is an increasingly appropriate and convenient industrial decontamination process. DIC reactors are currently operating at laboratory, pilot, and industrial scales. Thus, there are several infrastructure models with different features and capabilities; the energy consumption has been calculated to be 0.110 kWh per kg and per cycle.

This work described the identification and industrial uses of a thermomechanical destruction of microorganisms, which is mainly valid for solid products or powders. DIC can advantageously replace conventional processes in this field where many new treatments, such as radiation (γ, ultraviolet, acoustic) or mechanical (UHP, ultrasound) treatments, when possible, have only had limited use.

References

Allaf K, Debs-Louka E, Louka N, Abraham G (1998) Procédé de réduction ou d'élimination d'organismes, de microorganismes, de pasteurisation et de stérilisation des produits solides en morceaux ou pulvérulents et installation pour la mise en œuvre d'un tel procédé. 98/02032,

Allaf T, Besombes C, Mih I, Lefevre L, Allaf K (2011) Decontamination of solid and powder foodstuffs using DIC technology. In: InTech (ed) Advances in computer science and engineering. Matthias Schmidt, Croatia

Debs-Louka E, Louka N, Abraham G, Chabot V, Allaf K (1999) Effect of compressed carbon dioxide on microbial cell viability. Appl Environ Microbiol 65(2):626–631

Le Jean G, Abraham G, Debray E, Candau Y, Piar G (1994) Kinetics of thermal destruction of Bacillus stearothermophilus spores using a two reaction model. Food Microbiol 11(3):229–241, http://dx.doi.org/10.1006/fmic.1994.1026

Lin H-M (1984) Peng-Robinson equation of state for vapor–liquid equilibrium calculations for carbon dioxide + hydrocarbon mixtures. Fluid Phase Equilibria 16(2):151–169, http://dx.doi.org/10.1016/0378-3812(84)85028-1

Lin H-M, Chan E-C, Chen C, Chen L-F (1991) Disintegration of yeast cells by pressurized carbon dioxide. Biotechnol Prog 7(3):201–204. doi:10.1021/bp00009a001

Lin HM, Yang Z, Chen LF (1992) An improved method for disruption of microbial cells with pressurized carbon dioxide. Biotechnol Prog 8(2):165–166. doi:10.1021/bp00014a012

Part III
Extraction

Extraction of Essential Oils and Volatile Molecules

Baya Berka-Zougali, Colette Besombes, Tamara Allaf, and Karim Allaf

1 Introduction

By and large, essential oils are a volatile mixture of organic compounds derived from a single botanical source. These "oils," composed of distinctive components (sometimes hundreds of chemical aromas and compounds), are to a large extent responsible for the characteristic smell of the natural plant source. They are well characterized in terms of their aromatic uses, antimicrobial properties, and even in the field of aromatherapy.

Essential oils are usually extracted using hydrodistillation or steam distillation. These techniques have changed very little until recently. Heat is provided from boiling water or steam by both convection and/or condensation processes. Steam distillation includes a solid–gas interaction stage, whereas hydrodistillation involves a solid–liquid interaction. At the operational level, for both techniques, the vapors generated have to be condensed on an adequate low temperature condenser followed by decantation/separation between water and the other condensed liquids. Issues with the conventional extraction processes are linked to the low extraction kinetics which is known to be particularly slow (a long processing time of up to 24 h and sometimes longer). This involves consumption of a large amount of energy (Bocchio 1985; Cassel et al. 2009) and generates thermal degradation of some

B. Berka-Zougali
Laboratory of Research on Bio-active Products and Valorization of Biomasse, Ecole Normale Supérieure, Vieux-Kouba, 16050 Alger, Algeria

C. Besombes (✉) • K. Allaf
Laboratory of Engineering Science for Environment (LaSIE FRE 3474) CNRS, 17042 La Rochelle Cedex 01, University of La Rochelle, La Rochelle, France
e-mail: colette.besombes@univ-lr.fr; kallaf@univ-lr.fr

T. Allaf
ABCAR-DIC Process, La Rochelle, France

T. Allaf and K. Allaf (eds.), *Instant Controlled Pressure Drop (D.I.C.) in Food Processing: From Fundamental to Industrial Applications*, Food Engineering Series,
DOI 10.1007/978-1-4614-8669-5_6,

compounds in the extract (Jiménez-Carmona et al. 1999; Gámiz-Gracia and Luque de Castro 2000). Furthermore, other undesirable aspects are encountered while using conventional essential oil extraction methods, e.g., loss of some volatile compounds and low extraction efficiency (EE). To deal with these issues, new and innovative processes have been defined and developed (Allaf et al. 1993, 1998) such as instant controlled pressure drop (Détente Instantanée Contrôlée DIC) which is a solvent-free process that can be used to rapidly isolate high-quality essential oils.

1.1 Vapor Pressure and Boiling

Temperature is a parameter that expresses the mean fluctuating translation kinetic energy of molecules forming a thermodynamic system. The distribution of the kinetic energy of each molecule in both a liquid and vapor system provides an equilibrium between molecules leaving and entering the surface of the liquid. The vapor pressure depends only on the temperature and indicates the number of molecules that can be released into the surrounding medium with a high kinetic energy at the exchange surface.

Volatility characterizes the tendency of a substance to enter the vapor phase. Indeed, at a given temperature, a dynamic equilibrium between the liquid and vapor phases reflects the mean fluctuating kinetic energy. Volatility is a thermodynamic characteristic that is strictly linked to the composition and density of translation fluctuating kinetic energy, i.e., the temperature. However, the vapor pressure is not dependent on liquid concentration in the material. Indeed, the vapor pressure of any substance only depends on the temperature and is independent of the total external pressure. Generally speaking, substance volatility is defined in terms of the temperature; the higher the temperature, the higher the "concentration of vapor phase molecules" and the greater the volatility (Mortimer 2008). In other words, at a given temperature, the higher the vapor pressure of a substance, the more volatile it is.

The boiling temperature of a substance is thermodynamically defined as the temperature at which the vapor pressure is identical to the surrounding pressure. The definition of a compound's volatility should not and cannot be limited to the boiling temperature.

1.2 Impact in Terms of Distillation Processes

Hydrodistillation of volatile compounds can be performed to separate these compounds from a liquid mixture; this involves a solid–liquid interaction followed by a liquid–vapor interaction. Steam distillation is usually carried out by establishing an interaction between a solid matrix containing volatile compounds and steam as a vector of heat; this is a solid–liquid–vapor interaction. Both distillation processes depend on the differences in vapor pressures of the individual components at a given temperature. The relative concentration in the vapor phase is almost independent from the concentration in the liquid phase.

1.3 Relative Volatility

Relative volatility is a measure of the ratio of the volatilities of two compounds at a given temperature (Fig. 1). When relative volatility is very close to one, this indicates that they have very similar vapor pressure values. This means that it is not possible to separate the two components via distillation at this given temperature. Here relative volatility is defined relative to water vapor pressure.

$$\text{Relative volatility} = p_v(\text{substance})/p_v(\text{water}) \quad (1)$$

With p_v is the vapor pressure.

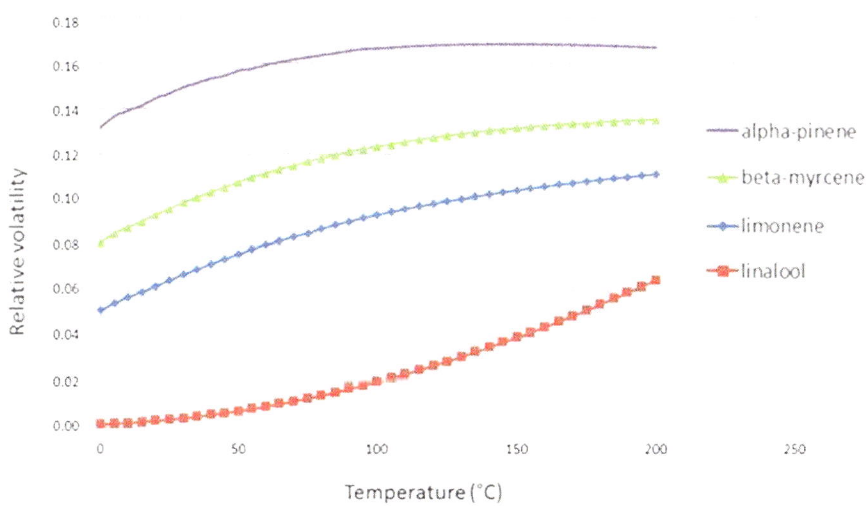

Fig. 1 Relative volatility of orange peel volatile compounds vs. water

2 Issues with Standard Essential Oil Extraction

In our studies of volatile molecules we mainly focused on essential oil extraction. Extraction issues can concern several aspects: plant structure, heat transfer, and mass (liquid and vapor) transfers.

Essential oils are usually extracted using hydrodistillation or steam distillation. These techniques provide heat from steam or boiling water by both convection and condensation.

- Steam extraction includes a solid–gas interaction stage
- Whereas hydrodistillation involves a solid–liquid followed by a liquid–vapor interaction

At the operational level, in both techniques, vapors (water and volatile molecules such as essential oils) generated at high temperature have to be condensed on an adequate low temperature condenser. This is followed by a decantation stage in which water is separated from the other condensed liquid. The kinetics of the evaporation stage of these techniques is known to be particularly slow (a long processing time of up to 24 h and sometimes longer), involving consumption of large amounts of energy and the degradation of sensitive molecules (Bocchio 1985; Gámiz-Gracia and Luque de Castro 2000; Cassel et al. 2009).

2.1 Surface Secretion: Heat and Fluid Surface Interaction

The essential oil extraction process depends on the way the plant stores oil and the type of secretion zones. There are special zones involved in the secretion of essential oils located on the surface of the plant (peltate glandular trichomes) or within plant tissues (deep structures).

Whatever the heat source (boiling water, saturated steam, etc.), if the zone is located on the surface, the essential oil extraction time is short as it only depends on the vapor pressure gradient of essential oils between the steam-plant exchange surface and the low temperature condenser.

Washing process: Essential oil (EO) and water vapor flow:

$$\dot{m}_{\mathrm{v,EO}} = \frac{\mathrm{d}m_{\mathrm{v,EO}}}{\mathrm{d}t} = k_{\mathrm{EO}}\, A\, [p_{\mathrm{EO}}(T_{\mathrm{X-s}}) - p_{\mathrm{EO}}(T_{\mathrm{cond}})] \quad (2)$$

$$\dot{m}_{\mathrm{v,W}} = \frac{\mathrm{d}m_{\mathrm{v,W}}}{\mathrm{d}t} = k_{\mathrm{W}}\, A\, [p_{\mathrm{W}}(T_{\mathrm{X-s}}) - p_{\mathrm{W}}(T_{\mathrm{cond}})] \quad (3)$$

Convection/condensation heat flow:

$$\dot{Q} = h\, A\, [T_{\mathrm{steam}} - T_{\mathrm{X-s}}] \quad (4)$$

Heat flow balance:

$$\dot{Q} = \dot{m}_{v,EO} L_{v,EO} + \dot{m}_{v,W} L_{v,W} \tag{5}$$

with $\dot{m}_{v,EO}$ and $\dot{m}_{v,W}$ are the essential oil vapor rate and the water vapor rate, respectively (kg s^{-1}); k the exchange coefficient of mass transfer by diffusion/convection in the surrounding medium (m^{-1} s or kg s^{-1} m^{-2} Pa^{-1}); A and $\langle A \rangle$ the effective exchange area at the solid-plant surface and the average of effective exchange area between the solid-plant surface and the condenser surface, respectively (m^2); (T_{X-s}), (T_{cond}), and (T_{steam}) the temperatures at the exchange surface, at the condenser surface, and of steam, respectively (°C); $p_{EO}(T_{X-s})$ and $p_{EO}(T_{cond})$ the vapor pressure of essential oils considered as one volatile substance, at exchange surface temperature (T_{X-s}), and at condenser temperature (T_{cond}), respectively (Pa); $\dot{Q}$ the total heat flow rate exchanged at the exchange surface (W); $L_{v,EO}$ and $L_{v,W}$ the evaporation latent heat for essential oils EO and water, respectively (J kg^{-1}); h the heat exchange coefficient by global convection/condensation processes (W m^{-2} K^{-1}).

Equation 5 indicates that most of the heat Q is absorbed and used for evaporating water and essential oils (with a ratio which is normally the relative volatility, as calculated above). However, while the separation step is immediate in steam distillation, the presence of essential oils in water in hydrodistillation means that an additional second stage would need to be added, which then increases the complexity of the operation.

For this type of peltate glandular trichome plant, as the secretion zones of essential oils are distributed on the surface, the process would usually be *intensified* by: (1) increasing the difference in temperature and reducing the distance between the exchange surfaces of the material and the condenser, respectively; and (2) by establishing a convection/agitation vapor transport.

Specific Walls of Secretion Organs at the Surface: Resistance to Thermal Treatment

The membrane and walls of the secretion organ may be a watertight, *impenetrable barrier to liquid and vapor*. The vapor pressure of essential oils and water at the external wall of the exchange surface, surface $p_{EO}(T_{X-s})$ and $p_W(T_{X-s})$, respectively, normally depend on temperature but are generally very weak.

Mass transfer through these membranes and walls becomes the limiting process. Direct mechanical breakage of the gland walls is the only effective way to recover these essential oils.

2.2 Deep Secretion Zones of Essential Oils

Where there is a homogeneous distribution of essential oil secretion zones in the deep structures, a phenomenological modeling of steam extraction can be made by considering both heat and mass transfers within the porous matrix, coupled to the external transport (Allaf et al. 2011).

2.2.1 Structural Characteristics

One of the most important characteristics of porous material is the *specific bulk density* ρ_{bulk}. This describes the apparent volume normally measured through the bouncy force (Archimedes method) or, for water-sensitive materials, using fine mineral powder such as Fontainebleau sand or cracking powder (80 μm diameter) (Louka and Allaf 2002, 2004).

The intrinsic or even true density $\rho_{\text{intrinsic}}$ is determined and used to identify the "porosity" and the absolute expansion ratio. The solid density $\rho_{\text{intrinsic}}$ is measured using the Pycnometer method, which employs Archimedes' principle of fluid (usually Helium) displacement and Boyle's law to determine the volume and density of the solid. The absolute expansion ratio can be quantified by determining the ratio between the intrinsic and the bulk (specific) densities:

Absolute expansion ratio:

$$\xi_{\text{abs}} = \frac{\text{Apparent volume}}{\text{Intrinsic volume}} = \frac{\rho_{\text{intrinsic}}}{\rho_{\text{bulk}}} \tag{6}$$

Porosity:

$$\text{Porosity ratio } \psi = \frac{\text{Apparent volume} - \text{Intrinsic volume}}{\text{Apparent volume}} = \frac{\xi_{\text{abs}} - 1}{\xi_{\text{abs}}} \tag{7}$$

Another important porous material parameter should normally be defined: the average size of the pores or the *mean pore diameter* δ:

$$\delta = f(r) \tag{8}$$

Pore diameter may play a very important role by intensifying heat flow, combining conduction with a phenomenon of evaporation–condensation at the pore wall. Similarly, heat transfer may occur by conduction with a higher effective conductivity.

2.2.2 Internal Transfers

Since the external mass transport from the surface to the surrounding environment is easily achieved, whatever the "volatile" compound, internal heat and mass transfer are the potential limiting processes of a given operation.

Internal Heat Transfer

Heat flow is used in part to increase the product temperature, but in the main it ensures a phase change (liquid–vapor). In steam extraction, the external saturated steam ensures heat transfer on the external surface of the solid matrix, which normally occurs through convection/condensation.

After quickly rising to the highest temperature (very close to the saturated steam temperature) the exchange surface can ensure a gradual spread of heat within the solid.

Subsequently, internal heat transfer occurs through a similar conduction phenomenon. The partial presence of volatile molecules within the porous structure should entail a phenomenon of *convection-vaporization/condensation in the holes*. This generates a heat transfer phenomenon where the temperature gradient is the driving force, with an effective conductivity value that is much higher than the standard conductivity values.

Internal Liquid and Vapor Mass Transfers

Typically, essential oil secretion cells are distributed uniformly in the internal volume. It is generally assumed that during steam extraction of volatile compounds, the transfer of liquid essential oil is negligible and that most mass transfer is a Fick-type gas diffusion, where the vapor pressure gradient of each volatile compound is the driving force.

2.2.3 Paradox of Coupled Heat and Vapor Transfers

From paragraph 0 on "Relative volatility," it is worth emphasizing the idea that the vapor pressure of each substance, either water or other volatile compounds, depends only on the temperature.

Mass transfer processes, which usually occur with liquids and vapors, depend on the nature and localization of the molecules and the porosity of the medium. These processes may be driven by capillary forces and by diffusion when it is a liquid/solid interaction. The gas phase (gas/solid interaction) mass transfer occurs through diffusion within the holes. It also depends on several factors including temperature, structure, and morphological state of the matter (porosity ratio, mean pore size and distribution, specific surface area, permeability of the secretion element walls, etc.) (Allaf 2009). In steam extraction, it can be assumed that the transfer of essential oils in the form of a liquid within the solid matrix is negligible. The main transfer phenomenon occurs in the gas diffusion phase in which the structure, the presence of secretion cavity walls as barriers, and more generally the porosity of the material all play a role. Macroscopic mass transfer inside the material is a Fick-type law related to the vapor pressure gradient of each volatile compound through an effective diffusivity (D_{eff}).

In similar cases, it is assumed that essential oil extraction necessarily implies an amount of heat capable of transforming the liquid phase into gas within the porous material (Besombes et al. 2010). Since the external vapor pressure is generally saturated, one can postulate:

$$\vec{\nabla}.\vec{\varphi}+\left(\rho_{\mathrm{s}}c_{\mathrm{ps}}+\rho_{\mathrm{EO}}c_{\mathrm{pEO}}+\rho_{\mathrm{w}}c_{\mathrm{pw}}\right)\frac{\partial T}{\partial t}+\frac{\partial}{\partial t}\left[\frac{\psi}{RT}(p_{\mathrm{EO}}M_{\mathrm{EO}}L_{\mathrm{EO}}+p_{\mathrm{w}}M_{\mathrm{w}}L_{\mathrm{w}})\right]=0 \quad (9)$$

As the internal transfer is carried out by conduction:

$$\vec{\nabla}.\left(-\lambda_{\mathrm{eff}}\vec{\nabla}T\right)+\left(\rho_{\mathrm{s}}c_{\mathrm{ps}}+\rho_{\mathrm{EO}}c_{\mathrm{pEO}}+\rho_{\mathrm{w}}c_{\mathrm{pw}}\right)\frac{\partial T}{\partial t}+\frac{\partial}{\partial t}\left[\frac{\psi}{RT}(p_{\mathrm{EO}}M_{\mathrm{EO}}L_{\mathrm{EO}}+p_{\mathrm{w}}M_{\mathrm{w}}L_{\mathrm{w}})\right]=0 \quad (10)$$

where φ is the heat flow within the porous material (W m^{-2}); ρ_{s} the apparent density of dry material (kg m^{-3}); c_{ps} the specific heat of dry material (J kg^{-1} K^{-1}); ρ_{eo} the apparent density of liquid essential oils in the material (kg m^{-3}); c_{pe} the specific heat of liquid essential oils in the material (J kg^{-1} K^{-1}); ρ_{w} the apparent density of water in the material (kg m^{-3}); c_{pw} the specific heat at a constant pressure of water (J kg^{-1} K^{-1}); T the temperature (K); t the time (s); ψ the porosity ratio; R the ideal gas constant (J mol^{-1} K^{-1}); p_{EO} the vapor pressure of essential oils in the material (Pa); M_{EO} the molar mass of essential oils (kg mol^{-1}); L_{EO} the latent heat of essential oil vaporization (J kg^{-1}); p_{w} the vapor pressure of water in the porous material (Pa); M_{w} the molar mass of water (kg mol^{-1}); L_{w} the latent heat of water vaporization (J kg^{-1}); λ_{eff} the effective conductivity of the porous wet material (J m^{-1} K^{-1}).

The temperature distribution is assumed to be stationary during most of the operation since in the main heat transfer is "only" used to evaporate essential oils and water (Allaf 2009):

$$\vec{\nabla}.\left(-\lambda_{\mathrm{eff}}\vec{\nabla}T\right)+\frac{\partial}{\partial t}\left[\frac{\psi}{RT}(p_{\mathrm{EO}}M_{\mathrm{EO}}L_{\mathrm{EO}}+p_{\mathrm{w}}M_{\mathrm{w}}L_{\mathrm{w}})\right]=0 \quad (11)$$

As the vapor mass transfers of different essential oil compounds and water are governed by a Fick-type law, the formulations of Allaf (1982) can be used separately:

$$\frac{(p_{EO}/T)}{\rho_s}\left(\vec{v}_{EO} - \vec{v}_s\right) = -D_{\mathrm{eff}_{EO}} \vec{\nabla}\left(\frac{p_{EO}/T}{\rho_s}\right) \tag{12}$$

$$\frac{(p_w/T)}{\rho_s}\left(\vec{v}_W - \vec{v}_s\right) = -D_{\mathrm{eff}_w} \vec{\nabla}\left(\frac{p_w/T}{\rho_s}\right) \tag{13}$$

v_{EO} is the absolute velocity of essential oils within the porous solid (m s^{-1}); v_s the absolute porous solid velocity (m s^{-1}); v_w the absolute velocity of the water vapor within the porous solid (m s^{-1}); $D_{\mathrm{eff}_{EO}}$ the effective diffusivity of essential oil within the porous solid (m^2 s^{-1}); D_{eff_w} the effective diffusivity of vapor within the porous solid (m^2 s^{-1}).

By neglecting possible shrinkage phenomena, one can assume that ρ_s = constant and $\vec{v}_s = 0$, (12) and (13) may be transformed into:

$$(p_{EO}/T)\vec{v}_{EO} = -D_{\mathrm{eff}_{EO}} \vec{\nabla}\,(p_{EO}/T) \tag{14}$$

$$(p_w/T)\vec{v}_W = -D_{\mathrm{eff}_w} \vec{\nabla}\,(p_w/T) \tag{15}$$

Since the external vapor pressure is saturated, it can be assumed that the internal water vapor pressure p_w in the porous material is approximately constant, and so (11) can be postulated as:

$$-\lambda_{\mathrm{eff}} \vec{\nabla} \cdot \left(\vec{\nabla}\, T\right) + \psi M_{EO} L_{EO} \frac{\partial}{\partial t}\left[\frac{p_{EO}}{RT}\right] = 0 \tag{16}$$

By assuming that each "particle" of the plant is spherical, homogeneous, and isotropic, (15) and (16) can be transformed in one dimension (r) into:

$$\frac{(p_{EO}/T)}{\rho_s} v_{EO} = -D_{\mathrm{eff}_{EO}}\left(\frac{\partial(p_{EO}/T)}{\partial r}\right) \tag{17}$$

$$-\lambda \frac{\partial^2 T}{\partial r^2} + \psi M_{EO} L_{EO} \frac{\partial}{\partial t}\left[\frac{p_{EO}}{RT}\right] = 0 \tag{18}$$

To achieve this kind of diffusion, the essential oil vapor flow should be directed from $r = 0$ towards $r = R$ (Fig. 2):

$$v_{EO} > 0 \tag{19}$$

$$\frac{\partial(p_{EO}/T)}{\partial r} < 0 \tag{20}$$

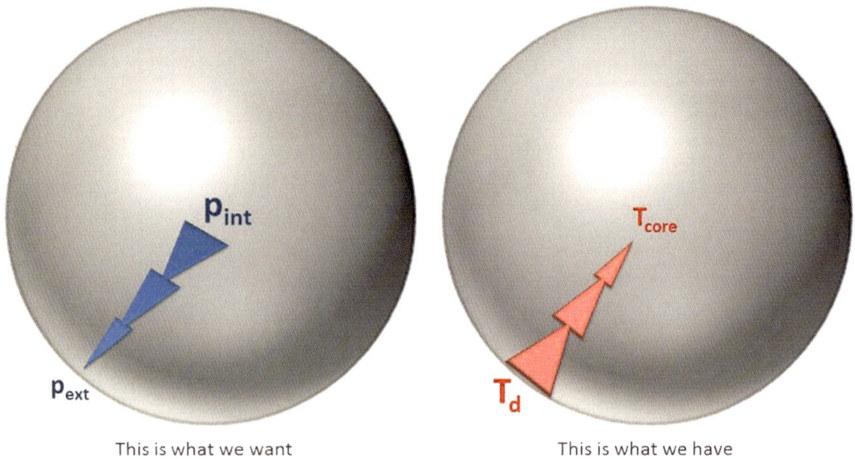

Fig. 2 Essential oil (*left*) and heat (*right*) transfers required for an extraction

However, as the exchange surface defined by $r = R$ has the highest temperature, and as the values of p_{EO}/T increase with increasing temperature, p_{EO}/T also has the highest value at $r = R$ (Fig. 3):

$$\frac{\partial(p_{EO}/T)}{\partial r} > 0 \tag{21}$$

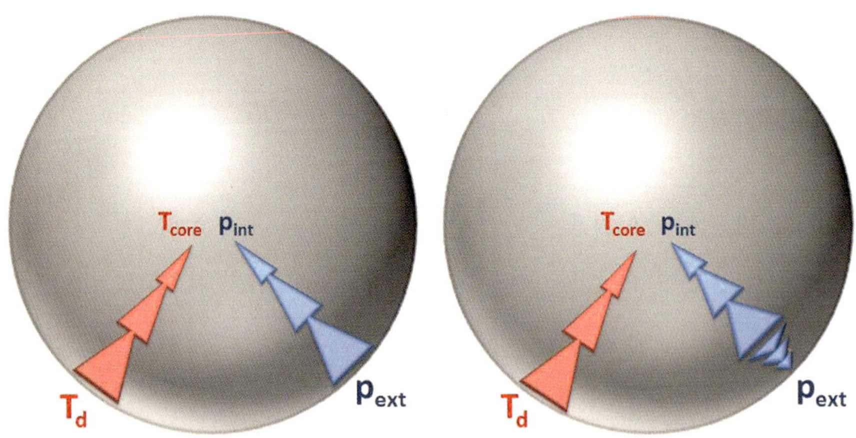

Fig. 3 The paradoxical situation of the extraction of volatile compounds during steam distillation; impact in terms of progressive front kinetics

As long as the essential oil vapor pressure p_{EO} is highest at the exchange surface, essential oil transfer occurs both towards the granule core and the surrounding medium. This results in a paradoxical situation, where part of the vapor is transferred towards the core, which is the complete opposite to what is required for an extraction operation (Al Haddad 2007; Al Haddad et al. 2008). It also means that steam extraction is achieved through "front progressive" kinetics (Fig. 3). Moreover, thermal degradation of the matrix might decrease the transfer because of structure shrinkage and increasing hardness.

This further explains why a pretreatment step to reduce granule size by grinding is often required in steam extraction. However, one real possibility to remedy this situation and notably improve the kinetics is to adopt another heating method such as microwaves or DIC. These achieve mass transfer through total pressure gradient (TPG) processes (Darcy's law).

3 Intensification of Essential Oil Extraction Using DIC

3.1 Principle of Instant Controlled Pressure Drop

The main issue in standard steam distillation, as mentioned above, could be identified as the paradoxical situation, i.e., the partial transfer of vapor towards the core of the porous plant. Instant controlled pressure drop technology (DIC: Détente Instantanée Controlée) overcomes this problem. Indeed, DIC treatment involves a Darcy-type transfer of volatiles from inside the material towards the surrounding medium, which is under a vacuum.

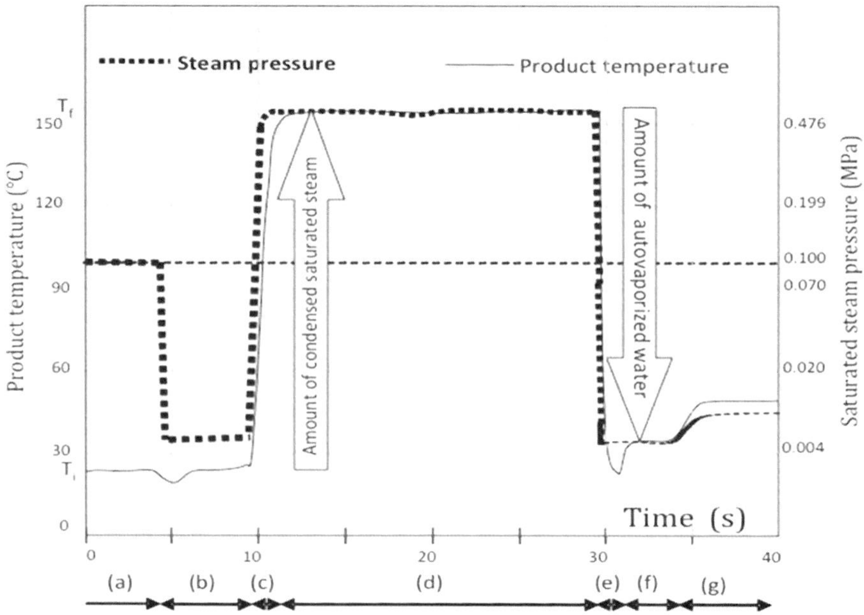

Fig. 4 Change in temperature and pressure during a DIC saturated steam pressure treatment: (*b*), (*c*) and (*d*) are the heating steps and (*e*) is the one-step pressure drop stage: (*a*) atmospheric pressure; (*b*) initial vacuum; (*c*) injection of saturated steam; (*d*) steam pressure and ambient temperature kept constant; (*e*) abrupt pressure drop towards a vacuum; (*f*) maintaining the vacuum; (*g*) release to atmospheric pressure

DIC treatment is based on two main stages of heating and a drop in pressure. The first stage is a high temperature/high pressure–short time effect. It is induced by exposing the raw material for a short period to high temperature and high pressure using one of various methods: saturated steam pressure, high temperature compressed air, high pressure and microwaves, etc. The pressure is generally between 0.1 and 1 MPa, depending on the product (i.e., a temperature from 100 to 180 °C in the case of saturated steam; Fig. 4). Hence, the vapor pressure of each essential oil compound is higher than that of standard steam distillation, which is carried out at 100 °C (the boiling temperature of water at atmospheric pressure).

The high temperature–short time stage is followed by an abrupt pressure drop towards a vacuum at about 5 kPa. This type of drop in pressure, at a rate ($\Delta P/\Delta t$) that is higher than 0.5 MPa s^{-1}, simultaneously triggers:

- An autovaporization of volatile compounds
- The instantaneous cooling of the product, which stops thermal degradation
- Swelling or even rupture of the cells and the secretion cavity walls

The porous structure thus created enhances mass transfer, intensifying the effective diffusivity as well as the overall operation kinetics. However, the major impact of DIC is to allow essential oil transfer along a gradient of total pressure (Darcy-type law), flowing from the core of the material at high total pressure of vapor to its surface under a vacuum.

The temperature and pressure levels during one DIC cycle are shown in Fig. 4. After an initial atmospheric pressure stage (a), a vacuum of about 4.5 kPa is established in the autoclave (b) and, just after, saturated steam is injected into the autoclave (c) to be maintained at a fixed pressure for a predetermined time (d). The initial vacuum (a) allows closer contact between the steam, which is the heating fluid, and the exchange surface of the plant. The vacuum results in a much larger fractal exchange surface. Indeed, at atmospheric pressure, the open and dead-end pores are normally filled with air, which acts as an isolating medium, resulting in very weak thermal conductivity and little contact with condensed water. Thus, the vacuum step increases heat transfer towards the exchange surface of the plant, and the product immediately reaches saturated steam temperature. After this thermal treatment, the steam is cut off and the spherical valve is opened rapidly (in less than 50 ms), which results in an abrupt pressure drop within the vessel (e). The vacuum period (f) is followed by a final release to atmospheric pressure (g) or by stage (c) of a new DIC cycle for a multi-cycle DIC treatment. Multi-cycle DIC contains n repetitions of stages (c–f), with the sum of heating times as the total heating time.

3.2 *Fundamental Approach in the Different Steps*

3.2.1 Heating Stage

The first DIC stage is the heating stage. The aim is to establish, from the initial product temperature T_i, a uniform treatment temperature T_t that is practically equal to the boiling temperature of water at the treatment pressure. Heating by pressured saturated steam injection is a specific process that begins by quickly bringing the product surface (exchange surface) to the steam temperature T_t, partly by convection but mainly by condensation. This is followed by a conduction process within the product. Inside a wet porous material, the possibility of having an evaporating/condensation process generally results in a much higher effective conductivity λ_{eff} than the calculated or measured static value λ.

To evaluate the quantity of water released after steam condensation, the heat energy needed to modify the temperature from the initial T_1 level to the treatment temperature value T_t must be calculated. When the initial wet porous material is defined as a dry material m_d with an initial water content W (% dry basis), it should be possible to consider the specific heat of the material as a linear combination of both compounds. At constant treatment pressure P_t, the enthalpy variation (or heat absorbed by the product) should be the following (Choi and Okos 1986):

Sensible heat Q at constant pressure:

$$c_\mathrm{p} = \sum_{i=1}^{n} \left(C_i c_{\mathrm{p},i} \right) \tag{22}$$

Specific heat at constant pressure:

$$c_\mathrm{p} = \frac{1}{m} \left[\frac{\partial Q}{\partial T} \right]_\mathrm{P} \tag{23}$$

In our case, this situation becomes:

Sensible heat Q at constant pressure:

$$\Delta H = Q = m_\mathrm{d} \left(c_{\mathrm{p,d}} + W c_{\mathrm{p,w}} \right) \left(T_\mathrm{t} - T_\mathrm{i} \right) \tag{24}$$

where m_d is the dry weight of the material; c_p the specific heat of the dry material, m_d, and water, W, at a constant pressure. It is expressed in J kg^{-1} K^{-1}.
The increasing amount of water from the condensation of steam at the product exchange surface becomes:

m_v:

$$m_\mathrm{v} = m_\mathrm{d} \frac{\left(c_{\mathrm{p,d}} + W c_{\mathrm{p,w}} \right) \left(T_\mathrm{t} - T_\mathrm{i} \right)}{L} \tag{25}$$

Just before dropping the pressure, there is an increase in water content (%, dry basis) in the product:

$$\Delta W = \frac{\left(c_{\mathrm{p,d}} + W c_{\mathrm{p,w}} \right) \left(T_\mathrm{t} - T_\mathrm{i} \right)}{L} \tag{26}$$

Since the heating fluid is saturated steam, the kinetics of the operation is strictly linked to the condensation, which is a very rapid phenomenon:

Condensation kinetics:

$$\frac{\mathrm{d}m_\mathrm{v}}{\mathrm{d}t} = kS \left(T_\mathrm{t} - T_\mathrm{s} \right) \tag{27}$$

By assuming convection is negligible in the heating process, the heat flow from the saturated steam towards the surface is considered to be:

Heating of surface kinetics:

$$\frac{dQ}{dt} = \dot{Q} = \phi = \varphi S = kLS(T_t - T_s) \tag{28}$$

where k is the coefficient of condensation, S the exchange surface between heat and the product (kg m^{-2} s^{-1} K^{-1}); L the latent heat of evaporation (J kg^{-1} = m^2 s^{-2}); T_s the temperature at the exchange surface going from T_i to T_t.

When steam is injected under initial atmospheric air pressure conditions, the open and dead-end pores remain filled with air, which acts as an insulating medium, so the effective exchange surface A is limited to the geometrical smooth surface. In contrast, after an initial vacuum step, the steam injected almost completely fills the superficial open and dead-end pores, leading to close contact between the steam and the pores. Thus, the effective fractal exchange surface A_{eff} between the condensed steam and the product is now much higher than the value A:

After vacuum step condensation kinetics:

$$\frac{dm_v}{dt} = kA_{eff}(T_t - T_s) \tag{29}$$

The condensation kinetics is much quicker when this type of initial vacuum step is used prior to the actual heating step. However, as the specific heat c_p of air is negligible, the amount of condensed steam m_v is the same in both cases.

This first washing step ensures an interaction with condensed steam at the surface that is strictly correlated with the temperature T_t. Water and heat diffusion transfer within the material occurs at the highest values of water content and temperature from the surface towards the core.

Since the surface temperature is established quickly at steam pressure T_t, thanks mainly to saturated steam condensation, heat transfer within the product starts immediately. Heat and mass transfers occur separately within the product. The first occurs through conduction according to a Fourier type law:

$$\vec{\varphi} = -\lambda_{eff}.\vec{\nabla}\,T \tag{30}$$

The second is a normal Fick-type water diffusion mass transfer (Allaf 1982):

First Fick's law (Allaf et al. 2011):

$$\frac{\rho_W}{\rho_d}\left(\vec{v}_W - \vec{v}_d\right) = -D_{eff}\vec{\nabla}\left(\frac{\rho_W}{\rho_d}\right) \tag{31}$$

Absence of expansion or shrinkage ($\vec{v}_d = 0$ and $\rho_d =$ constant):

$$\rho_W \vec{v}_W = -D_{eff} \vec{\nabla} \rho_W \tag{32}$$

By adding the continuity:

$$\frac{\partial \rho_W}{\partial t} = -\vec{\nabla} \cdot \left(D_{eff} \vec{\nabla} \rho_W \right) \tag{33}$$

where λ_{eff} is the effective conductivity involving the phenomenon of evaporation/condensation within the porous material (W m^{-1} K^{-1}); ρ_W the apparent density of water within the material (kg m^{-3}); ρ_d the apparent density of the dry material (kg m^{-3}); $\vec{v}_W$ the absolute velocity of water within the material (m s^{-1}); $\vec{v}_d$ the absolute velocity of the material (m s^{-1}), normally due to the expansion or the shrinkage; D_{eff} the effective water diffusivity within the dry material (m^2 s^{-1}).

By adding the continuity:

$$\frac{\partial T}{\partial t} = -\vec{\nabla} \cdot \left(\alpha_{eff} \vec{\nabla} \rho_W \right) \tag{34}$$

The values of both effective diffusivities α_{eff} and D_{eff} expressed in m^2 s^{-1} depend on water content W (dry basis). We usually use the Dickerson approximation:

Dickerson (1969) for estimation of effective thermal diffusivity α (m^2 s^{-1}):

$$\alpha_{eff} = \frac{8.8\ 10^{-8} + W\alpha_W}{1 + W} \tag{35}$$

Effective thermal diffusivity α_{eff} depends to a small degree on the temperature T, the porosity ratio, and the average size of the pores, with values about 10^{-7} m^2 s^{-1}. In contrast, the effective diffusivity of water within the product D_{eff} varies considerably with the porosity and temperature of the granule with values (about 10^{-10} m^2 s^{-1}) that are generally assumed to be much lower than α_{eff}. The temperature within the product reaches its equilibrium level very quickly compared to the time needed to attain the equilibrium level of water content. The D_{eff} value is assumed to be constant if the hypothesis of both structural and thermal homogeneities is accepted. A general quantification of physical processes and adequate experiments confirms this hypothesis, so (33) becomes Fick's second law:

Similar second Fick's law:

$$\frac{\partial \rho_w}{\partial t} = D_{eff} \vec{\nabla} \left[\vec{\nabla} \rho_w \right] = D_{eff} \nabla^2 \rho_w \tag{36}$$

By assuming a unidirectional flow, (36) becomes:

One-r-direction second Fick's law:

$$\frac{\partial \rho_{\mathrm{w}}}{\partial t} = D_{\mathrm{eff}} \frac{\partial^2 \rho_{\mathrm{w}}}{\partial r^2} \tag{37}$$

The duration (t_c) of the high temperature stage of DIC must generally be defined so that both the temperature (T) and moisture content (W) are uniformly distributed within the product. Since the former is achieved much more quickly, the duration of the DIC high temperature stage is calculated to obtain a uniformly deep distribution within the product of the moisture condensed on the exchange surface.

3.2.2 Pressure Drop Stage

The pressure drop stage has to be defined and studied at several levels: autovaporization, expansion, cooling, and glass transition processes.

By abruptly dropping the pressure, an adiabatic autovaporization of the overheated water and volatile compounds occurs, inducing an instant cooling of the residual material.

The vapor generated by this process engenders mechanical stresses and constraints within the plant, swelling both cells and the structure. The modification in general behavior and structure greatly depends on the plant's viscoelastic characteristics, which are a function of temperature and water content. However, the preservation of the new swelled structure depends greatly on the glass transition T_g, which occurs at low temperature and water content. This aspect is very important and explains why the pressure has to drop towards a vacuum allowing the equilibrium temperature to reach T_g level. The amount of water removed from the product by autovaporization through the exchange surface becomes:

m_v:

$$m_{\mathrm{v}} = m_{\mathrm{d}} \frac{\left(c_{\mathrm{p,d}} + W' c_{\mathrm{p,w}}\right)\left(T_{\mathrm{p}} - T_{\mathrm{i}}\right)}{L} \tag{38}$$

where T_p is the temperature reached by the product just after the pressure drops.

Depending on the duration of the pressure drop, T_p can be the equilibrium level T_e (quasi-static transformation):

For a quasi-static, long decompression:

$$T_{\mathrm{p}} = T_{\mathrm{e}} \tag{39}$$

T_p can also be much lower than the equilibrium temperature T_e when the pressure is dropped instantaneously (instantaneous transformation):

For instantaneous pressure drop:

$$T_p \le T_e \tag{40}$$

Thus, just after the pressure is dropped, the water content (%, dry basis) of the product decreases:

Modification of water content (% dry basis):

$$\Delta W = \frac{\left(c_{p,d} + W' c_{p,w}\right)\left(T_p - T_t\right)}{L} < 0 \tag{41}$$

where W' is the initial water content increased by the amount condensed during heating (26). After the instant pressure drop step (e) and during the vacuum step (f), the total internal pressure is mainly due to water vapor (and other volatile molecules) in the porous medium. Under DIC conditions, Allaf and colleagues assumed that the total pressure of the volatile mixture vapor in the porous material just after the pressure drop was much higher than the external pressure (Allaf 2009).

3.2.3 Mass Transfer by TPG

In various cases where there is a zone of TPG (membrane transfer, permeability of porous material), a Darcy mass transfer occurs, with the TPG as the "driving force":

General Darcy law:

$$\rho_v\left(\vec{v}_v - \vec{v}_d\right) = -\frac{K}{\nu_v}\vec{\nabla} P \tag{42}$$

where the permeability K expressed in (m^2) mainly depends on the structure, porosity, and tortuosity of the solid matrix. The kinematic viscosity ν_v expressed in ($m^2\ s^{-1}$) depends on the nature of the transferred fluid (density, molecular size). Allaf (2009) assumed that an instantaneous expansion occurs, possibly due to the autovaporization itself just after the pressure drops. Sulaiman (2013) found the expansion time to be between 20 and 200 ms, depending on the material and the autovaporization conditions. By assuming that $\vec{v}_d = 0$ and the transfer to be a one-dimension radial within a spherical shaped material, (42) becomes:

$$\rho_v \vec{v}_v = -\frac{K}{\nu_m}\frac{\partial P}{\partial r}\vec{e}_r \tag{43}$$

Therefore, the transfer of volatile molecule vapor such as water and essential oils within the expanded granule is assumed to be from the core towards the surrounding medium along the gradient of total pressures. By using the first Darcy-type law describing this transfer, coupled to mass conservation and continuity, and by integrating between a pore (whose radius is R_o) and the external radius (R_s) of the spherical shape, (43) becomes:

$$\dot{m}_v = \frac{4\pi K}{\nu_m} \frac{(P_{globule} - P_{ext})}{\left(\frac{1}{R_o} - \frac{1}{R_s}\right)} \tag{44}$$

The total pressure ($P_{globule}$) in the hole decreases with time (t), depending on the flow ($\dot{m}_m$). The globule radius (R_o) is assumed to be constant (absence of expansion) as is the temperature, which would appear to remain constant at the value reached just after the pressure drop, defined by the water vapor/temperature/pressure equilibrium. Equation 44 becomes:

$$\frac{dP_{globule}}{dt} \frac{MV_{hole}}{RT} = -\frac{4\pi K}{\nu_v} \frac{(P_{globule} - P_{ext})}{\left(\frac{1}{R_o} - \frac{1}{R_s}\right)} \tag{45}$$

$$(P_{globule} - P_{ext}) = (P_{globule,o} - P_{ext}) \exp\left(-\frac{3KRT}{\nu_v} \frac{t}{MR_o^3\left(\frac{1}{R_o} - \frac{1}{R_s}\right)}\right) \tag{46}$$

The change in the amount of vapor with time is:

$$\dot{m}_v = \dot{m}_{v,o} \exp\left(-\frac{3KRT}{\nu_m} \frac{t}{MR_o^3\left(\frac{1}{R_o} - \frac{1}{R_s}\right)}\right) \tag{47}$$

Characteristic time for Darcy's mass transfer:

$$\tau = \frac{MR_o^3\left(\frac{1}{R_o} - \frac{1}{R_s}\right)\nu_m}{3KRT} \tag{48}$$

Generally, the duration t_v of the vacuum stage (f) is sufficient to allow the vapor mixture to be transported towards the surrounding medium and then collected.

The DIC process provides an appropriate solution to the problems posed by the paradoxical phenomenon and greatly intensifies essential oil extraction. It uses autovaporization instead of (or coupled with) the evaporation process, and TPG instead of the molecular diffusion phenomenon. Here too, the water vapor pressure plays a very important role as the driving force because it is generally ten times higher than the vapor pressure of the natural volatile molecules. Another point which has to be highlighted is that by grinding the material and reducing the granulometry, the characteristic time τ of Darcy's mass transfer (which evolves mainly as a function of R_o^2) can be significantly decreased.

3.3 Experimental Work

Various raw materials were used, as well as different extraction processes and assessment methods. Hydrodistillation and steam distillation were used as references for a comparison with DIC instant autovaporization. The main part of our experimental work was carried out at different scales for analyzing, optimizing, and modeling. DIC essential oil extraction was studied with saturated steam pressure, total heating time, and the number of cycles as the operative parameters (independent variables). The response parameters we had to consider as dependent variables were the total yield and the variation in extract composition.

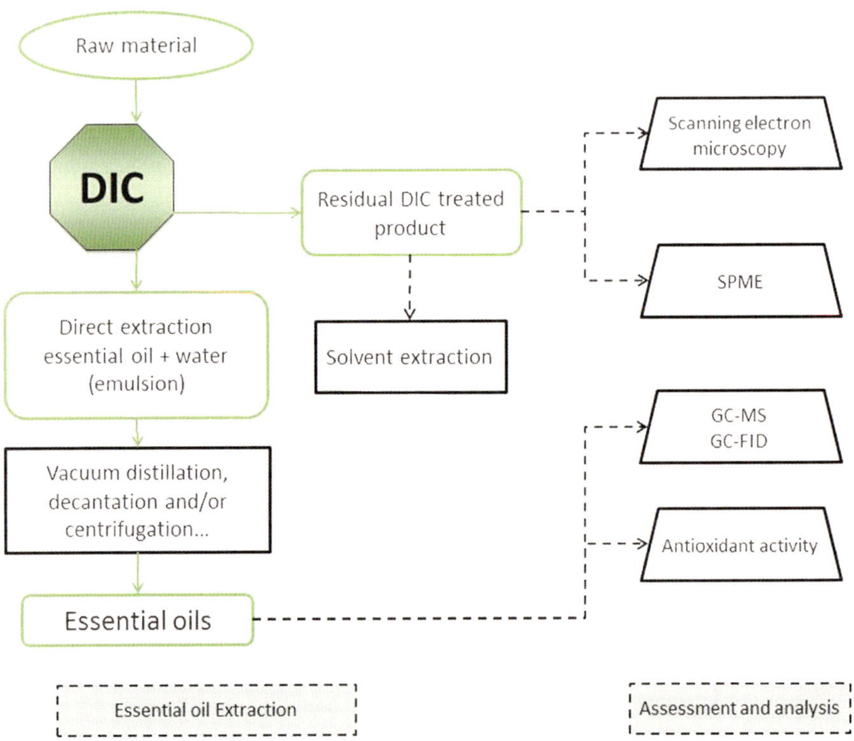

Fig. 5 DIC treatment protocol

3.3.1 Protocol and Products

DIC extractions were carried out on numerous raw materials, such as angelica seeds, cinnamon bark, chamomile flowers, lemongrass leaves, patchouli leaves, thyme leaves, valerian roots, citrus peels, lavandin, and myrtle, and the results compared with those obtained using conventional extraction methods.

The different materials were studied using the protocol shown on the diagram above (Fig. 5). Moreover, the operating parameters were analyzed using the experimental designs.

3.3.2 Standard Treatment Methods

Steam distillation and hydrodistillation were used as a reference for a comparison with DIC essential oil extraction.

Steam Distillation

A Pignat P3734 laboratory apparatus was used. A packed bed of a variable quantity (approximately 50 g) of aromatic raw material was placed on a stainless steel grill fixed above a distillation vessel containing 2 L of water. Electric heating of the water produced 2.5 kg h^{-1} of steam, which in turn passed through the plant bed, where it evaporated and carried away the essential oil. The vapors were condensed and then collected in a receiver vessel.

Hydrodistillation

With most of the products, the hydrodistillation equipment used was a modified Clevenger. In accordance with conventional hydrodistillation protocols, 200 g of raw material was immersed in 2 L of distilled water in a 3 L distillation flask. The extraction of essential oils was carried out from the first drop of distillate until the plant material had been completely consumed.

Measurements and Assessments

The main measurements and assessments were devoted to optimizing the operating parameters in order to obtain the highest quality of end product and the best operation performances, including yields and extraction efficiency, kinetics, fractioning capacity, composition, and antioxidant activity.

$$\mathrm{EE}\,(\%) = \frac{\text{DIC extract EO}}{\text{Raw material EO}} \tag{49}$$

The measurement of residual essential oils in DIC textured material provided the extraction efficiency using the extraction ratio:

$$\text{Extraction ratio}\,(\%) = \frac{\text{Raw material EO} - \text{DIC textured sample EO}}{\text{Raw material EO}} \tag{50}$$

DIC extraction efficiency (EE) should correlate with the essential oil extraction ratio. However, it is noteworthy that DIC treatment isolates part of the compounds from the residual solid and usually increases their availability. EE may then be higher than 100 %.

All DIC extraction operations were compared with standard extraction methods, whether hydrodistillation or steam distillation.

Optimized DIC extraction systematically provided a higher or equal yield of essential oils compared to hydrodistillation or steam distillation. Moreover the duration of the extraction was much shorter with DIC instant autovaporization. Hence, compared to hydrodistillation, DIC extraction required 4 min instead of 3 h for lavandin, 3 min instead of 4 h for rosemary, and 2 min instead of 3 h for myrtle (Table 1).

Table 1 Yield of essential oils obtained using DIC extraction in comparison with conventional hydrodistillation

		Lavandin	Rosemary	Myrtle
	Yields (g EO/100 g dm)	4.3	1.3	0.57
Hydrodistillation	Treatment time	180 min	240 min	180 min
DIC extraction	Treatment time	4 min	3 min	2 min
References		Besombes et al. (2010)	Allaf et al. (2013b)	Berka-Zougali et al. (2010)

The total yield of total essential oils extracted from various aromatic plants was measured using DIC extraction and compared with conventional steam distillation and hydrodistillation. Table 2 presents the results obtained, with thyme as an example.

Table 2 Yield of essential oils from thyme leaves obtained by DIC extraction in comparison with conventional steam distillation and hydrodistillation

Essential oil yields in thyme leaves	Steam distillation	Hydrodistillation	DIC-extraction
(g EO/100 g dm)	2.71	3.68	4.34

More severe DIC conditions were needed to isolate the essential oil from cananga flowers. Steam distillation took 12 h, whereas Kristiawan et al. (2008) found that a yield of 2.77 % of dry matter was obtained with eight DIC cycles at 0.6 MPa in 4 min, compared with 24 h with hydrodistillation (Table 3).

Table 3 Yield of essential oils (g EO/100 g dm) from cananga flowers obtained by DIC extraction in comparison with steam distillation

	Steam distillation	DIC 1-extraction	DIC 2-extraction
Essential oil yields (%)	2.71	3.68	4.34
Treatment time	24 h	1 min	4 min

The DIC yield of total essential oils was, in most cases, slightly higher than with hydrodistillation and steam distillation (Table 4).

Table 4 Yield of essential oils from orange peel obtained by DIC extraction in comparison with conventional hydrodistillation (Allaf et al. 2013a)

		Orange peel	
Essential oils		In pieces	Ground
Hydrodistillation	Yields (g EO/100 g dm)	0.20 ± 0.01	1.61 ± 0.05
	Extraction time	5 h	3 h
DIC extraction	Yields (g EO/100 g dm)	1.64 ± 0.05	*No need*
	Extraction time	3 min	

Autovaporization vs. Evaporation

In order to define whether direct DIC extraction occurs through an evaporation or autovaporization effect, the total heating time (t) and the number of cycles (C) were prioritized through a statistical study. The main objective was to identify separately the impacts of evaporation and autovaporization through macroscale modeling. Indeed, as DIC involves high temperatures (here up to 160 °C), the vapor pressure must be much higher than that of conventional steam extraction, which is usually performed at 100 °C. Nevertheless, fundamental and experimental studies have already proved that autovaporization, which is strictly linked to the pressure drop, is the major factor in the extraction of essential oils by DIC (Besombes et al. 2010; Allaf et al. 2011).

A lot of experimental work has demonstrated this fundamental observation. Indeed, a 2-parameter, 5-level experimental design was used with different products. DIC operating parameters were the total thermal treatment time t and the number of cycles C, which is the number of pressure drops (Table 5). t represents the evaporation process while C delineates the effect of autovaporization. Obviously, essential oil extraction efficiency (EE), defined by (49), was selected as the response dependent variable. An appropriate statistical treatment of the data was used to identify the most significant factor.

Table 5 Operating parameters (independent variables) and ranges used in response surface methodology (RSM) at a fixed steam pressure with dried aromatic plants

	Coded level	$-\alpha$	-1	0	$+1$	$+\alpha$
$X1 = C$	Number of cycles	1	2	5	8	9
$X2 = t$	Heating time (s)	80	103	160	217	240
	Time of 1 cycle (s)	80	51.5	32	27.0	26.7

α (axial distance) $= \sqrt[4]{2^N}$, N is the number of independent variables. In the present case: $N = 2$ and $\alpha = 1.4142$

Numerous natural products were used (valerian root, lemongrass, chamomile, angelica seeds, thyme, lavandin, etc.) for essential oil extraction using DIC. A Pareto chart systematically confirmed that the number of cycles (which delineates the impact of autovaporization) is the only operating factor to have a significant effect on total essential oil extraction by DIC (Fig. 6).

To summarize, it is well established that essential oil extraction by steam distillation occurs through evaporation and is strictly dependent on processing time.

In contrast, direct DIC extraction is an autovaporization process during which evaporation barely occurs. Thus total thermal treatment time t has an insignificant impact in DIC. The only requirement is to define the duration of each cycle such that it is longer than the critical heating time that results in a uniform distribution of temperature and moisture content (5–15 s) throughout the product, and the vacuum time to enable the removal of the essential oil vapor according to Darcy's law, which is TPG (3–10 s).

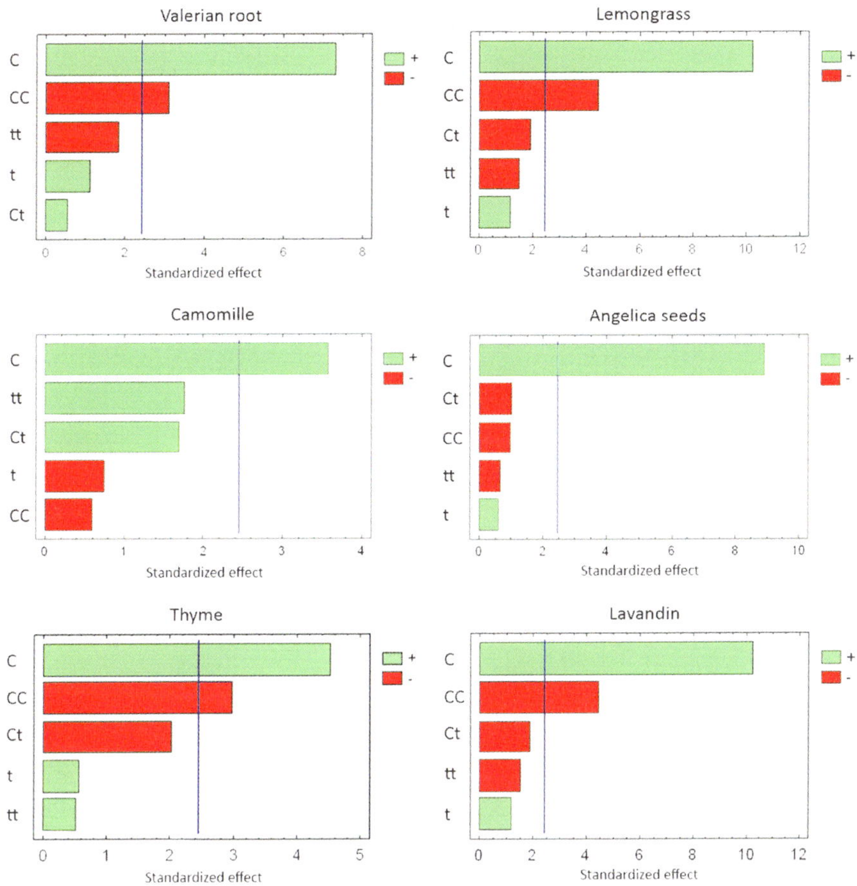

Fig. 6 Pareto charts: effect of *C* (number of cycles) and *t* (total processing time) on DIC extraction efficiency with valerian root, lemongrass, chamomile, angelica seeds, thyme, and lavender essential oils

Impact on Quality

Increasing the Proportion of Oxygenated Compounds

Since thermal treatment time can be perfectly controlled and is short enough to allow DIC to be a high temperature/short time (HTST) process, the quality of the essential oils extracted is preserved without any notable thermal degradation. Thus, in the case of Algerian myrtle leaves, a comparison of the composition of oils extracted by hydrodistillation or DIC proved that oxygenated compounds were largely preserved (Table 6). Similar results were systematically obtained when the total thermal treatment time was shorter than a certain limit (<2/4 min per cycle).

Table 6 Chemical composition of essential oil extracted from Algerian myrtle leaves by hydrodistillation or DIC

Compound family	HD-EO[a] (%)	DIC-EO[b] (%)
Hydrocarbon monoterpenes	55.17	26.15
Oxygenated monoterpenes	32.9	42.66
Benzenoid compounds	2.33	10.14
Hydrocarbon sesquiterpenes	4.76	9.79
Oxygenated sesquiterpenes	1.68	8.46
Total aliphatic compounds	0.5	0.34
Total non-oxygenated compounds	59.93	35.94
Total oxygenated compounds	37.41	61.6
Total identified compounds	97.34	97.54
Yield g EO/100 g dry matter	0.51 ± 0.04	0.56 ± 0.12

[a]HD-EO: Essential oils extracted by hydrodistillation from Algerian myrtle leaves
[b]DIC-EO: Essential oils directly extracted by DIC treatment from Algerian myrtle leaves

Another important point is that the antioxidant activity is a significant functional quality. The amount of DIC-EO needed to obtain the same level of antioxidant effect was systematically lower than with hydrodistillation-EO, i.e., DIC essential oils have a slightly superior antioxidant activity compared to those obtained by hydrodistillation. In the case of Algerian myrtle leaves the improvement was about 6 %.

Fractioning Extraction

In DIC treatment, each compound has a specific thermomechanical behavior in relation to autovaporization. This generally depends on its localization, its volatility which is governed by processing temperature and, possibly, the rate of the pressure drop. In DIC, it is easy to establish a relative fractioning of various compounds by (1) modifying the pressure/temperature, (2) separating compounds versus cycles, and (3) using sequenced condensers at different temperatures. Since the volatility of each compound is calculated from the vapor pressure at various temperatures, DIC fractioning extraction is simple if based on these extremely important thermodynamic characteristics, when they are available in the literature. The change in extraction rate versus time and DIC cycles is then simply monitored.

Indeed, in the case of lavandin essential oil, compounds extracted after each DIC cycle were determined as shown in Fig. 7. Some compounds such as limonene [$C_{10}H_{16}$], eucalyptol [$C_{10}H_{18}O$], ethyl linalool [$C_{12}H_{22}O$], and α-humulene [$C_{15}H_{24}$] were completely extracted after the first cycle. After the sixth cycle most compounds had been extracted. Only some, such as α-terpineol [$C_{10}H_{18}O$], τ-cadinol [$C_{15}H_{26}O$], and α-bisabolol [$C_{15}H_{26}O$], required many more pressure drops.

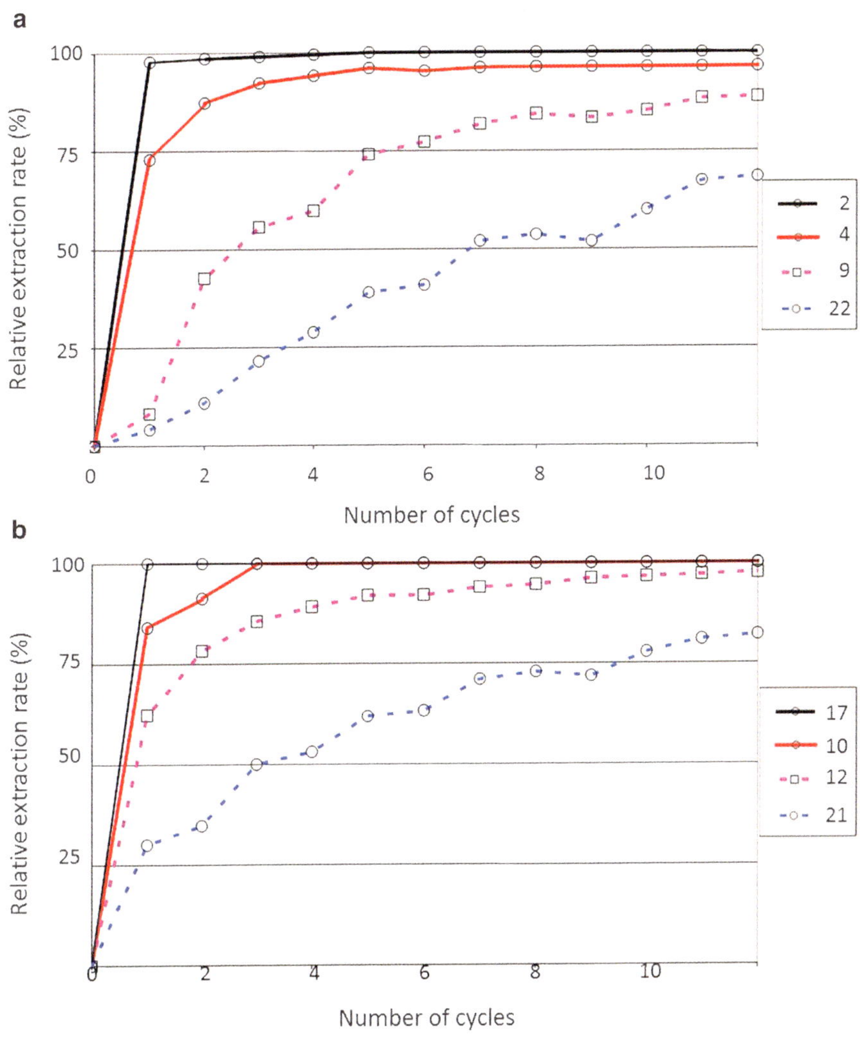

Fig. 7 Relative extraction rate of some lavandin essential oil compounds

This observation is very important in the strategy of DIC fractioning extraction. For each case, the process is automated through the definition of operating parameters and the design of DIC/condenser equipment depending on the raw material and the desired compounds. In other words, fractioning can be carried out during both the extraction and the separation phases.

3.4 Essential Oil-in-Water Emulsion

The extraction of essential oils by instantaneous autovaporization inexorably leads to an instant cooling of both EO and water vapors. A highly stable essential oil-in-water emulsion is recovered in the vacuum and condensation tank. Some specific uses of such emulsions can be defined. However, a second quasi-static "conventional" operation of evaporation/condensation can be inserted as a stage to separate the compounds of essential oils from the aromatic water.

3.5 Possible Recovery of the Residual Material

DIC pretreatment facilitates the solvent extraction unit operation by markedly increasing the availability of nonvolatile active molecules. Since DIC is a high temperature–short time (a few dozen seconds) process, it involves only weak thermal degradation and, similarly, DIC products have a much better technological aptitude for solvent extraction. DIC conditions are usually defined to simultaneously obtain the best possible extraction of essential oil and expansion of the raw material. The expanded residual solid is then recycled using various methods: drying, leading to expanded granule powder or, more especially, extraction with a solvent (water, ethanol, methanol, hexane, super-critical fluids, etc.). In both cases, DIC pretreatment facilitates the operation by significantly increasing the availability of nonvolatile active molecules. Since DIC is a high temperature–short time (a few dozen seconds) process, it involves only weak thermal degradation and, similarly, DIC products have a much better technological aptitude for solvent extraction.

4 Energy Consumption

While hydrodistillation conventionally requires 4–8 kg of added water with 0.7 kg of vapor/kg RM (raw material), DIC treatment needs about 0.125 kg of steam/kg RM per cycle, a total of 0.662 kWh kg^{-1} RM. The calculated energy consumption is 0.110 kWh kg^{-1} RM per cycle, i.e., 0.6 kWh kg^{-1} RM (Besombes et al. 2010; Allaf et al. 2011) while in hydrodistillation it is about 1.4 kWh kg^{-1} RM.

5 Conclusion

Although steam distillation (SD) is the most widely used essential oil extraction process, this tedious, time- and energy-consuming operation has unacceptable impacts in terms of yields, quality of essential oils, energy consumption, environmental impact, and recovering final residual materials. These different well-established drawbacks are becoming too serious given modern industrial needs and consumer requirements, and this has led to the definition of a new intensification scenario. Usually, the steam distillation process is limited by the diffusion of essential oil vapor within the product. However, the kinetics is much weaker than that imposed by Fick's diffusion because of the paradoxical situation. Thus, DIC treatment is a highly appropriate technology that can be used to remedy this situation and intensify the kinetics. After a short stage of saturated steam heating, the instantaneous pressure drop (DIC) process involves a very effective Darcy-type vapor transfer. This leads to a relevant, short time, high yield, quality controlled essential oil extraction. For instance, in the case of thyme oil, a direct oil yield of 4.3 % dm was obtained using 9 cycles, 160 s, 0.5 MPa in DIC compared with 3.5 % dm for a 7 h hydrodistillation (HD) and 2.7 % for a 6 h SD.

Work performed on a large number of raw materials showed that DIC extraction increased the extraction yield. Conventional extraction treatments such as steam distillation or hydrodistillation, do not achieve the same levels of performance.

DIC extraction increases the amount of essential oil obtained and reduces the duration of treatment. It simultaneously improves the kinetics and the yields.

For all aromatic plants studied, DIC extraction was more efficient, and the number of cycles was the most significant parameter. The experimental results corroborate the fundamental study: DIC extraction is achieved mainly through autovaporization; the contribution of "conventional" evaporation to this type of extraction process is negligible. The mass transfer is not controlled by the molecular diffusion of volatile components but by the TPG. Moreover, DIC has a second effect: vapor instantaneously generated in the flowers breaks and alveolates the structure and frees the internal content of the cells. The risk of thermal degradation is avoided by using a short heating time (<1 min) in each DIC cycle.

References

Al Haddad M (2007) Contribution théorique et modélisation des phénomènes instantanée dans les opérations d'autovaporisation et de déshydratation. Université de La Rochelle, La Rochelle, France

Al Haddad M, Mounir S, Sobolik V, Allaf K (2008) Fruits & vegetables drying combining hot air, DIC technology and microwaves. Int J Food Eng 4(6)

Allaf K (1982) Transfer phenomena and industrial applications. Lebanese University Faculty of Science, Beirut

Allaf K (2009) The new instant controlled pressure—drop DIC technology. In: Chemat F (ed) Essential oils and aromas: green extraction and application. Har Krishan Bhalla and Sons, New Delhi, pp 85–121

Allaf K, Louka N, Bouvier JM, Parent F, Forget M (1993) Procédé de traitement de produits biologiques en vue de la modification de leur texture, installations pour la mise en oeuvre d'un tel procédé et produits ainsi réalisés. 93/09728

Allaf K, Debs-Louka E, Louka N, Abraham G (1998) Procédé de réduction ou d'élimination d'organismes, de microorganismes, de pasteurisation et de stérilisation des produits solides en morceaux ou pulvérulents et installation pour la mise en œuvre d'un tel procédé. 98/02032

Allaf K, Besombes C, Berka-Zougali B, Kristiawan M, Sobolik V, Allaf T (2011) Instant controlled pressure drop technology in plant extraction processes. In: Lebovka N, Vorobiev E, Chemat F (eds) Enhancing extraction processes in the food industry, Contemporary food engineering series. CRC Press Taylor & Francis Group, Dublin, Ireland, pp 255–302

Allaf T, Tomao V, Besombes C, Chemat F (2013a) Thermal and mechanical intensification of essential oil extraction from orange peel via instant autovaporization. Chemical Eng Process. doi:http://dx.doi.org/10.1016/j.cep.2013.06.005

Allaf T, Tomao V, Ruiz K, Bachari K, ElMaataoui M, Chemat F (2013b) Deodorization by instant controlled pressure drop autovaporization of rosemary leaves prior to solvent extraction of antioxidants. LWT-Food Sci Technol 51:111–119. doi:http://dx.doi.org/10.1016/j.lwt.2012.11.007

Berka-Zougali B, Hassani A, Besombes C, Allaf K (2010) Extraction of essential oils from Algerian myrtle leaves using instant controlled pressure drop technology. J Chromatogr A 1217(40):6134–6142. doi:10.1016/j.chroma.2010.07.080

Besombes C, Berka-Zougali B, Allaf K (2010) Instant controlled pressure drop extraction of lavandin essential oils: fundamentals and experimental studies. J Chromatogr A 1217(44):6807–6815. doi:10.1016/j.chroma.2010.08.050

Bocchio E (1985) Hydrodistillation of essential oils: theory and applications, vol 63. Parfums, cosme´tiques, aro^mes. Socie ´te´ d'expansion technique et e´conomique, Parise

Cassel E, Vargas RMF, Martinez N, Lorenzo D, Dellacassa E (2009) Steam distillation modeling for essential oil extraction process. Ind Crops Prod 29(1):171–176. doi:10.1016/j.indcrop.2008.04.017

Choi Y, Okos M (1986) Effects of temperature and composition on the thermal properties of food. Food engineering and process applications. Elsevier Applied Science, London

Dickerson RW (1969) Thermal properties of foods. In: Tressler DK, Van Arsdel WB, Copley MJ (eds) The freezing preservation of food, 4th edn. AVI Publishing Company, Westport, CT, pp 26–51

Gámiz-Gracia L, Luque de Castro MD (2000) Continuous subcritical water extraction of medicinal plant essential oil: comparison with conventional techniques. Talanta 51(6):1179–1185, doi:http://dx.doi.org/10.1016/S0039-9140(00)00294-0

Jiménez-Carmona MM, Ubera JL, Luque de Castro MD (1999) Comparison of continuous subcritical water extraction and hydrodistillation of marjoram essential oil. J Chromatogr A 855(2):625–632, doi:http://dx.doi.org/10.1016/S0021-9673(99)00703-7

Kristiawan M, Sobolik V, Allaf K (2008) Isolation of Indonesian cananga oil using multi-cycle pressure drop process. J Chromatogr A 1192(2):306–318. doi:10.1016/j.chroma.2008.03.068

Louka N, Allaf K (2002) New process for texturizing partially dehydrated biological products using controlled sudden decompression to the vacuum: application on potatoes. J Food Sci 67(8):3033–3038

Louka N, Allaf K (2004) Expansion ratio and color improvement of dried vegetables texturized by a new process "controlled sudden decompression to the vacuum": application to potatoes, carrots and onions. J Food Eng 65(2):233–243. doi:10.1016/j.jfoodeng.2004.01.020

Mortimer RG (2008) Physical chemistry, 3rd edn. Elsevier Academic Press, Canada

Sulaiman I (2013) Fundamental and experimental approaches of processes of autovaporization and texturing of overheated materials using instant controlled pressure drop towards a vacuum. La Rochelle University, La Rochelle, France

DIC Texturing for Solvent Extraction

Tamara Allaf, Baya Berka Zougali, Cuong Van Nguyen, Mohamed Negm, and Karim Allaf

1 Introduction

There is currently great interest accorded to the extraction and use of natural molecules from plants and other organic materials in food, nutraceuticals, cosmetics, ingredients, flavors, and aromas. The pharmaceutical industry has also increased its demand for natural compounds in preference to synthetic molecules. Moreover, consumer demand for these kinds of products is a major driving force behind this strategic trend. Thus, scientists are attempting to adapt extraction processes to meet these strategic requirements in order to replace synthetic molecules by natural compounds.

Meeting this demand requires a study of the intensification and optimization of existing extraction processes in order to be able to deal with large varieties of molecules, which are sometimes present at very low concentrations or are difficult to extract. Thus, extraction kinetics and yields have to be greatly improved.

T. Allaf (✉)
ABCAR-DIC Process, La Rochelle, France
e-mail: tamara.allaf@gmail.com

B.B. Zougali
Laboratory of Research on Bio-active Products and Valorization of Biomasse, Ecole Normale Supérieure, Alger, Algeria

C.V. Nguyen
College of Engineering Technology, Can Tho University, Can Tho, Vietnam

M. Negm
Department of Special Food and Nutrition, ARC, Food Technology Research Institute, Giza, Egypt

K. Allaf
Laboratory of Engineering Science for Environment (LaSIE FRE 3474) CNRS, 17042 La Rochelle Cedex 01, University of La Rochelle, La Rochelle, France
e-mail: kallaf@univ-lr.fr

T. Allaf and K. Allaf (eds.), *Instant Controlled Pressure Drop (D.I.C.) in Food Processing: From Fundamental to Industrial Applications*, Food Engineering Series, DOI 10.1007/978-1-4614-8669-5_7,

Solvent extraction is a very important unit operation in the chemical, biochemical, food, cosmetics, and pharmaceutical industries. Solvent extraction is generally used to recover compounds from either a solid or a liquid. The sample is placed in contact with a solvent that dissolves the desired solutes.

The solvents used for the separation of plant materials are usually water, organic solvents, petrochemical solvents, and super- or subcritical fluids. Standard extraction techniques include several methods that all consist in a liquid solvent interacting with a solid material in order to dissolve the desired components (Leybros and Frémeaux 1990). However, in the case of plants, the solid–liquid extraction process essentially depends on the morphology of the plant material.

The limiting factor in conventional solvent extraction operations is often the slow diffusion of both the solvent through the solid matrix and the solute from the core to the surface (Allaf et al. 2011). To avoid this difficulty or reduce its impact, several solutions have been proposed as pretreatments: reducing the size of particles by crushing; breaking cell walls by enzyme treatment or using ultrasound; microwaves; accelerated solvent extraction or a special solvent such as a supercritical fluid. These options have not definitively resolved this issue.

Indeed, adequate texturing is an indispensable pretreatment that overcomes this problem. Instant controlled pressure drop (DIC) technology is an appropriate way to improve the internal structure of matter while preserving its chemical composition, even vis-à-vis heat-sensitive molecules. DIC-assisted solvent extraction of ingredients from botanical sources can be studied and its transfer from academic research into manageable solutions for industrial use achieved in a very short time. This type of pressure drop, at a rate ($\Delta P/\Delta t$) which is higher than 0.5 MPa s^{-1}, simultaneously triggers an autovaporization of volatile compounds, which is very effective for deodorization, an instantaneous cooling of the product, which stops thermal degradation, and swelling or even rupture of the cell walls. The resulting porous structure enhances mass transfer and intensifies both effective diffusivity and starting accessibility, thereby increasing the overall kinetics. The impact of DIC treatment on the extraction kinetics and on the overall performance is significant. First, manufacturing new equipment for the unit operation could dovetail with an approach combining fundamental analysis with rigorous process modeling, and empirical studies using experimental designs. Secondly, a close collaboration between scientists from the disciplines of botany, thermodynamics, analytical chemistry, mechanics, and process engineering would be the most appropriate strategy to adopt.

2 Solvent Extraction of Active Molecules

A solvent is chosen for its capability to "solubilize" one or more components from a solid or a liquid, thus generating an extract solution (solvent + solute). Extraction capacity and selectivity are both very important characteristics. Solvent extraction/separation of plant-based substances is, from a technological point of view, a solid/liquid interaction. The solvent acts as a carrier fluid (liquid) to bring specific molecules out of the solid.

The transport of these active molecules to the surroundings occurs through a diffusion whose driving force is the concentration gradient of solute between the inner solution near the solid phase (more concentrated) and the surrounding liquid phase. At the end of the operation, the system tends towards internal/external equilibrium and diffusion is close to zero. In contrast, if the liquid phase is continuously renewed, the diffusion continues until the complete exhaustion of the solid.

2.1 Identification of the Limiting Process

The solid/liquid interaction is achieved by close contact between the solid and the solvent. Since the solvent is well defined, it can be assumed that the dissolution is carried out instantaneously, once the solvent interacts with the solid. During the extraction, the concentration of solute in the solid varies continuously, which explains the non-stationary mass transfer.

The operation starts with an interaction on the surface, which is an external process of washing, and the solute is immediately transported within the surrounding solvent medium.

After this first stage, a succession of processes occur, reflecting the interaction in the volume between the solid and the solvent, provoking the separation of the molecules concerned (solute). These successive processes include:

1. Solvent diffusion within the solid matrix.
2. Internal solute dissolution in the solvent: this occurs within the solvent, which has diffused within the porous solid plant matrix (internal process).
3. Solute diffusion in the solvent within the solid matrix towards the surface. This can be considered as the diffusion of solute within the filled-with-solvent pores. This specific transport is a Fick-type diffusion process.
4. External convection and/or diffusion transfer of the solute from the surface of the solid to the external environment.

Thus, the extraction can be studied on a phenomenological basis in terms of kinetics and total yields. The kinetics is usually expressed in terms of total solute concentration in the solid per unit time (dX/dt). The sequential nature of these processes results in the operation kinetics following the slowest, limiting phenomenon. An intensification of the unit operation needs to overcome the process identified as limiting.

2.1.1 Dissolution of the Molecule To Be Extracted in the Liquid Solvent

Dissolution is the operation of dissolving a substance in a solvent. Dissolution tends towards thermodynamic equilibrium in terms of concentration and is described through the distribution, or equilibrium partitioning constant or coefficient $\bar{\omega}_e$, which is a function of the maximum solubility (saturation level), thereby leading to an equilibrium concentration between the extract and the dry matter (Schwartzberg and Chao 1982).

The higher the value of "$\bar{\omega}_e$", the more easily the compound dissolves in the solvent. The value of "$\bar{\omega}_e$" depends on the characteristics of the solvent, the compound to be extracted, and the temperature.

2.1.2 Transport of Dissolved Solute from the Surface Towards the Outside

Since the solvent is chosen to provide a very fast (instantaneous) dissolution of the molecules of interest, solute close to the exchange surface is transferred to the surrounding medium via diffusion and/or convection.

In the absence of any agitation, it is possible that the limiting process of the extraction is the transfer of solute from the solid surface towards the external solvent. To obtain a proper intensification of this external transfer, a convective rather than a diffusional transfer has to be established. This has to be achieved through adequate agitation of the external "solvent medium." The external transfer resistance then becomes negligible.

Since the appropriate solvent instantaneously dissolves the solute, the extraction operation begins with the solvent washing the surface, which is a very rapid phenomenon:

Extraction rate of solute from surface:

$$\frac{\mathrm{d}m_{\mathrm{s}}}{\mathrm{d}t} = k_{\mathrm{e}}\mathrm{ESA}(\bar{\omega}_{\mathrm{e}} - \bar{\omega}_{\mathrm{solvent}}) \tag{1}$$

where m_s is the mass of solute extracted by interaction between the solvent and the surface of the product (kg of solute); k_e the coefficient of dissolution of solute by the solvent (kg of solvent m^{-2} s^{-1}); ESA the exchange surface area between the solvent and the product (m^2); $\bar{\omega}_e$ the dissolving coefficient at equilibrium (kg of solute per kg of solvent); $\bar{\omega}_{\mathrm{solvent}}$ the solute dissolved in the solvent (kg of solute per kg of solvent).

The density of the extraction rate of solute from the surface per unit of dry matter:

$$\frac{\mathrm{d}X_{\mathrm{s}}}{\mathrm{d}t} = k_{\mathrm{e}}\mathrm{SESA}(\bar{\omega}_{\mathrm{e}} - \bar{\omega}_{\mathrm{solvent}}) \tag{2}$$

X_s is the density of solute per unit of dry matter extracted by interaction between the solvent and the surface of the product (kg of solute per kg of dry matter or % db); SESA the specific exchange surface area between the solvent and the product per unit of dry basis ($m^2\ kg^{-1}$).

By intensifying the convection via a sufficiently strong agitation of the solvent, the dissolution coefficient of the solute by the solvent (k_e) is increased. This amplifies the kinetics of the process compared to a simple diffusion process. An appropriate ultrasound treatment can also help increase the internal fluctuation motion of the particles and the solute flow from the surface towards the external medium.

Extraction processes can be enhanced by grinding the raw material—the smaller the average granule size, the higher the specific surface exchange area (SESA).

Another effective intensification method can be obtained by injecting solvent after an initial vacuum step. The open and dead-end pores are normally filled with air, which acts as an insulating medium. However, an initial vacuum stage prior to the injection of solvent would modify the initial situation and almost completely fill these pores. Thanks to the initial vacuum step there is close contact and exchange between the "total surface" and the solvent, thus avoiding the limitation of the geometrically smooth surface. The effective exchange surface area between the solvent and the product is now much higher, with a fractal value. After the vacuum step, the density of the extraction rate of solute from the surface per unit of dry matter becomes:

$$dX_s = k_e \mathrm{SESA}_{\mathrm{eff}}(\bar{\omega}_e - \bar{\omega}_{\mathrm{solvent}})dt \tag{3}$$

"External" intensification processes are generally easy to carry out. Once the external transfer phenomenon is perfectly intensified, it is no longer the limiting process. Solute extraction kinetics from the surface can then reach its highest level.

The total effect can be shown through the "starting accessibility" parameter:

Starting accessibility:

$$\delta X_s = k_e \mathrm{SESA}_{\mathrm{eff}}(\bar{\omega}_e - \bar{\omega}_{\mathrm{solvent}})\delta t \tag{4}$$

Usually, in kinetics studies, the starting accessibility δX_s does not involve any diffusion source or effect; it is the density of solute per unit of dry matter extracted by the initial interaction between the solvent and the surface of the product, which is expressed as kg of solute per kg of dry matter or % db. The lower the granulometry, the higher the starting accessibility.

2.1.3 Internal Transfer by Diffusion

After this first washing step to establish a solvent interaction at the surface, solvent diffusion transfer in the matrix occurs from the surface, where there is the highest solvent concentration, towards the core, which has the lowest concentration.

The diffusional solvent transfer within the product starts immediately. The potential initial vacuum state, the low granulometry after grinding, or the use of strong agitation only modify the starting accessibility δX_s. They do not affect the effective diffusivity value, but do reduce the time frame needed for the solvent to replace air in the interstitial and pore volume in the matrix. Solvent transfer within the plant structure (solid/liquid interaction) has as its driving force the gradient of apparent densities ratio of solvent to solid.

A second internal mass transfer is the solute-in-solvent transfer. This correlates with the random diffusion of solute molecules along the concentration gradient of the apparent density ratio of solute to solvent.

As the former is generally much quicker, most of the kinetics is accounted for by the solute-in-solvent transfer within the holes of the solid matrix, which is highly dependent on the porosity and tortuosity of the material.

Since this deep mass transfer process cannot be intensified through external mechanical or thermal changes, diffusion (Fick's-type) becomes the major transport phenomenon occurring within the solid matrix (Aguilera and Stanley 1999). Only a modification of shape (by grinding), an expansion improving the porosity, and/or ultrasonic treatment to increase the apparent agitation of molecules in the interstitial and pore volume can improve the process.

2.1.4 Kinetics Analysis

As mentioned just above, in any solvent extraction from plants (a porous solid material), a first solvent exchange surface interaction, or washing (Fernández et al. 2012), takes place over a short time frame. The solute immediately dissolves in the external solvent, to be removed from the surface to the surrounding solvent medium, generally by convection (Ben Amor and Allaf 2009; Amor et al. 2008; Ben Amor 2008). The amount is given by the "starting accessibility" δX_s (expressed in g of extract per g of dry material).

Subsequently, most of the operation is controlled through various processes of penetration of the solvent within the material (capillarity, molecular diffusivity, etc.). In such conditions, the driving force for solvent transfer is the gradient of apparent densities ratio of solvent to solid, with $D_{\mathrm{eff,s\text{-}L}}$ ($\mathrm{m^2\ s^{-1}}$) as the effective diffusivity (Allaf et al. 2011, 2012):

$$\frac{\rho_{\mathrm{solvent}}}{\rho_{\mathrm{d}}}\left(\vec{v}_{\mathrm{solvent}}-\vec{v}_{\mathrm{d}}\right)=-D_{\mathrm{eff,\,s-L}}\vec{\nabla}\left(\frac{\rho_{\mathrm{solvent}}}{\rho_{\mathrm{d}}}\right) \tag{5}$$

Because of the absence of expansion: $\vec{v}_{\mathrm{d}}=0$ and $\rho_{\mathrm{d}}=$ constant:

$$\rho_{\mathrm{solvent}}\vec{v}_{\mathrm{solvent}}=-D_{\mathrm{eff}}\vec{\nabla}\rho_{\mathrm{solvent}} \tag{6}$$

By adding the continuity:

$$\frac{\partial \rho_{\text{solvent}}}{\partial t} = -\vec{\nabla} \cdot \left(D_{\text{eff}} \vec{\nabla} \rho_{\text{solvent}} \right) \tag{7}$$

Starting with the short-time stage of solvent diffusion and then much more markedly with the solvent-filled structure, the solute is transferred inside the matrix according to a Fick's-type Law (Allaf et al. 2011) with the gradient of the apparent density ratio of solute to solvent as the driving force:

$$\frac{\rho_{\text{solute}}}{\rho_{\text{solvent}}} \left(\vec{v}_{\text{solute}} - \vec{v}_{\text{solvent}} \right) = -D_{\text{eff}} \vec{\nabla} \left(\frac{\rho_{\text{solute}}}{\rho_{\text{solvent}}} \right) \tag{8}$$

However, as the operation occurs in a presumably motionless solvent within the solid matrix ($\vec{v}_{\text{solvent}} = 0$), it is usually assumed that the amount of solute is very low and $\rho_{\text{solvent}} = \text{constant}$:

$$\rho_{\text{solute}} \vec{v}_{\text{solute}} = -D_{\text{eff}} \vec{\nabla} \rho_{\text{solute}} \tag{9}$$

Using the mass balance, we obtain:

$$\frac{\partial \rho_{\text{solute}}}{\partial t} = -\vec{\nabla} \cdot \left(D_{\text{eff}} \vec{\nabla} \rho_{\text{solute}} \right) \tag{10}$$

Although the effective diffusivity (D_{eff}) varies considerably with the temperature in the system, it can be considered constant based on the thermal homogeneity hypothesis, in which case the equation becomes (Mounir and Allaf 2008):

Constant D_{eff} because of temperature homogeneity:

$$\frac{\partial \rho_{\text{solute}}}{\partial t} = -D_{\text{eff}} \nabla^2 \rho_{\text{solute}} \tag{11}$$

And, for one-direction radial flow, it becomes:

$$\frac{\partial \rho_{\text{solute}}}{\partial t} = -D_{\text{eff}} \frac{\partial^2 \rho_{\text{solute}}}{\partial r^2} \tag{12}$$

with ρ_{solute} is the apparent density of the solute within the solid matrix (kg m^{-3}); ρ_{d} the apparent density of the solid dry material (kg m^{-3}); v_{solute} the velocity of the solute (m s^{-1}); v_{d} the velocity of the solid dry material (m s^{-1}); d_{p} the average radius; X the amount of solute extracted at time t (g g^{-1} dry material); X_{∞} the amount of solute within the matrix (g g^{-1} dry material); δX_{s} the starting accessibility (g g^{-1} dry material).

The solutions required for this diffusion equation are highly dependent on the initial and boundary solvent conditions.

Experimental data should be utilized to identify δX_s and D_{eff}. For the diffusion model, the standard Crank's solutions, which are a function of the geometry of the products, may be adopted (Crank 1975). However, the experimental data to be used to identify D_{eff} must exclude those close to the initial time and so standard Crank's solutions should only concern the time $t > t_o$ (Mounir and Allaf 2009):

As an example, Crank's solution for a sphere with a radius d_p is:

$$\frac{X_\infty - X}{X_\infty - X_{t_o}} = \sum_{i=1}^{\infty} \frac{6}{i^2\pi^2} \exp\left(-\frac{i^2\pi^2 D_{eff}}{{d_p}^2}(t - t_o)\right) \tag{13}$$

$$\frac{X_\infty - X}{X_\infty - X_{t_o}} = \frac{6}{\pi^2}\exp\left(-\frac{\pi^2 D_{eff}}{d_p^2}(t - t_o)\right) + \frac{6}{4\pi^2}\exp\left(-\frac{4\pi^2 D_{eff}}{d_p^2}(t - t_o)\right)$$
$$+ \frac{6}{9\pi^2}\exp\left(-\frac{9\pi^2 D_{eff}}{d_p^2}(t - t_o)\right) + \cdots$$

Crank's calculation to identify D_{eff} (first-term approach):

$$\frac{X_\infty - X}{X_\infty - X_{t_o}} = A\exp(-k(t - t_o)) \tag{14}$$

First-term approach:

$$\ln\left(\frac{X_\infty - X}{X_\infty - X_{t_o}}\right) = -k(t - t_o) \tag{15}$$

$$D_{eff} = k\frac{d_p^2}{\pi^2} \tag{16}$$

The starting accessibility δX_s is then calculated by extrapolating the diffusion model to $t = 0$; $X_o \neq (X_i = 0)$:

$$\delta X_s = X_o - X_i = X_o \tag{17}$$

2.2 *Intensification of the Process*

2.2.1 Mass and Heat Diffusivities

A bibliographic analysis indicated that the effective diffusivity (D_{eff}) of the entire extraction operation for liquids in plant matrices is of a magnitude that is generally close to 10^{-11} to 10^{-10} m^2 s^{-1}, while heat diffusivity is usually between 10^{-8} and 10^{-6} m^2 s^{-1}, depending on water content and porosity.

This confirms, in a way, that isothermal solvent diffusion into a solid matrix is often the slowest, limiting process throughout the operation. The intensification operation lies in the expansion of the natural plant structure as an appropriate method for improving the ability to process the plant using solvent extraction.

2.2.2 Identification of the Extraction Operation

A kinetics model of solvent extraction can be used to quantify the effects of an intensification procedure via yields X_∞, starting accessibility (δX_s), and effective diffusivity (D_{eff}). The higher the values of X_∞, δX_s, and D_{eff}, the greater the plant's "technological aptitude" for solvent extraction technology.

Four procedures capable of intensifying solvent extraction have been described (Mounir and Allaf 2008):

1. Agitation of the surrounding solvent normally allows the transport of solute from the product surface to the outside by convection. By adding an initial vacuum step, the effective exchange surface area attains a much higher value than the smooth surface alone. Both operations involve a higher δX_s.
2. Grinding normally reduces the depth of internal diffusion in the granule. It also increases the surface exchange area and makes the solute on the surface more accessible. It results in a higher δX_s.
3. The aim of texturing is to reduce mass transfer resistance. Indeed, the natural plant structure and more specifically the cytoplasmic membrane and cell walls tend to restrain liquid transfer processes. Hence, using DIC swelling, and breaking membrane walls using enzymatic treatment, etc., the effective diffusivity D_{eff} mass transfer can be significantly increased.
4. Ultrasonic treatment (US) affects the exchange between the solvent and the surface of the material, which should increase the starting accessibility δX_s. Inside the porous material, the solute dissolved in the solvent normally has to flow into the pores according to Fick's law of diffusion. Here too, US involves a higher effective diffusivity D_{eff} through an authentic mass convection instead of natural diffusion.

It is important to highlight the fact that in all these intensification methods, the role played by instant controlled pressure drop (DIC) technology is strategic. Indeed in all the solvent extraction operations and whatever the part of plant, it is possible to intensify the operation through expansion, which increases the porosity, the specific surface area, and the availability of the desired compounds. Combined with grinding, it provides expanded granule powders with a much higher exchange surface. By establishing a vacuum stage prior to solvent injection and US treatment, a bona fide synergy of intensification is obtained.

2.3 *Instant Controlled Pressure Drop*

2.3.1 DIC: Texturing

Instant controlled pressure drop (DIC) consists in establishing a short time frame (from 10 s^{-1} min) high-temperature/high-pressure steam (0.05–1 MPa) step, followed by an instant pressure drop towards a vacuum (up to 1.5 kPa) with a decompression ratio $(\Delta P/\Delta t) > 0.2$ MPa s^{-1}. This involves an autovaporization of the volatile molecules (present in the solid matrix), thus inducing a cooling effect and texturing. In addition, the decompression towards a vacuum provides a greater temperature difference between the two states just before and after decompression. The amount of vapor produced is easily sufficient to develop the mechanical constraints without reaching an excessively high processing temperature (Iguedjtal et al. 2008; Louka and Allaf 2004), with a greater likelihood of achieving glass transition.

Following theoretical work on the thermomechanical expansion of organic polymers (Allaf 1988) and studies on the instantaneous transformations (Allaf 2002), DIC has been used in many different fields (Allaf and Vidal 1989; Allaf et al. 1993, 1998) and has been the subject of several Ph.D. theses on drying (Sahyoun 1996), drying/texturing (Ben Amor 2008), steaming (Habba 1997), drying fish (Juhel 2000), decontamination (Debs-Louka 2000), treatment of archaeological wood (Sanya 2000), drying of pharmaceutical products (Rakotozafy 2001), sputtering (Delgado-Rosas 2002), treatment of vegetables and leguminous plants (Haddad 2002), processing rice (Thai-Cong 2003), and the new concept of expanded granule powders (Mounir 2007). Various industrial applications have shown that DIC provides both a high product quality and appropriate process performances, thus saving energy and preserving the environment (green process).

2.3.2 Gas Pressure Texturing

Recent research on DIC has shown that it is possible to use gases such as air, carbon dioxide, and nitrous oxide at high pressures of 0.1–10 MPa. Changes in texture were observed, while other mechanisms (micromechanical, chemical) have still not been fully exploited with particularly heat-sensitive materials such as dairy powders.

2.3.3 Fundamental Studies and Modeling

Autovaporization

The product is initially placed at a temperature that is much higher than the final thermal equilibrium level, and simply by exposing it to an instant pressure drop, there is an abrupt release of a certain quantity of vapor. The volume of "vapor" is capable of inducing high amplitude internal constraints:

Change in water content dry basis by autovaporization:

$$\Delta W = \frac{m_{\mathrm{v}}}{m_{\mathrm{d}}} = \frac{\left(c_{\mathrm{p,\,d}} + W_{\mathrm{i}} c_{\mathrm{p,\,W}}\right)\left(T_{\mathrm{e}} - T_{\mathrm{t}}\right)}{L} < 0 \tag{18}$$

where m_{v} is the mass of vapor generated by increasing the treatment pressure (initial pressure) to the final pressure; m_{d} the mass of dry matter in the treated product; W_{i} the initial moisture content of the treated product as a % of dry matter; $c_{\mathrm{p,d}}$ and $c_{\mathrm{p,W}}$ the specific heat capacities of the dry matter and water, respectively; L the latent heat of evaporation of water; T_{t} and T_{e} the treatment temperatures before and after the pressure drop, respectively.

Normal volume of generated vapor:

$$V_{\mathrm{v}} = \frac{m_{\mathrm{s}} \Delta W}{M_{\mathrm{W}}} \frac{T_{\mathrm{e}}}{P_{\mathrm{f}}} = \frac{m_{\mathrm{s}}}{M_{\mathrm{W}}} \frac{R_{\mathrm{GP}} T_{\mathrm{e}}}{P_{\mathrm{f}}} \frac{\left(c_{\mathrm{p,\,s}} + W_{\mathrm{i}} c_{\mathrm{p,\,W}}\right)\left(T_{\mathrm{t}} - T_{\mathrm{e}}\right)}{L} \tag{19}$$

where M_{W} and ρ_{W} are the molar weight and the density of water, respectively; P_{f} is the final pressure (surrounding pressure) just after the pressure drops; R_{GP} is the ideal gas constant.

Assuming that the volume occupied by the vapor just when it is generated is determined by V_{a}, which is the difference between the specific volume and the intrinsic volume of the material, we can write:

Volume initially occupied by water:

$$V_{\mathrm{a}} = m_{\mathrm{s}} \left(\frac{1}{\rho_{sp\acute{e}cific}} - \frac{1}{\rho_{\mathrm{intrinsic}}} \right) \tag{20}$$

Constraints arising from the vapor generated should correlate with the difference between internal and external pressures:

$$\Delta P_{\mathrm{max}} = \frac{m_{\mathrm{s}} |\Delta W|}{M_{\mathrm{W}}} \frac{R_{GP} T_e}{V_a} - P_{\mathrm{f}} = \frac{R_{GP} T_e \left(c_{p,\,s} + W_{\mathrm{i}} c_{p,\,W}\right)\left(T_t - T_e\right)}{M_{\mathrm{W}} \left(\frac{1}{\rho_{specific}} - \frac{1}{\rho_{intrinsic}} \right) L} - P_{\mathrm{f}} \tag{21}$$

The maximum pressure difference $\Delta P_{\max}$ between the pore and the external environment is a very important parameter in texturing operations, whether they may be extrusion-cooking, steam-explosion, puffing, or DIC. Indeed, the pressure difference ΔP is the driving force in the change in the pore size with time:

Change with time of pore radius:

$$\mu\, 2\pi R \frac{\mathrm{d}R}{\mathrm{d}t} = \frac{1}{2}\pi R^2 \Delta P \tag{22}$$

with R is the radius of the pore; μ the apparent viscosity of the matrix at the temperature and humidity considered.

The rate of change of pore size is thus proportional to the pressure difference ΔP between the inside of the pore and the external medium, and is inversely proportional to viscosity. However, as this pressure difference depends on the initial value and changes according to the pore size, we can write a relation similar to that used by Arhaliass et al. (2009).

The rheological behavior of the material during expansion is very important. Indeed, the expansion ratio can be established by coupling autovaporized water, which is responsible for the internal constraints ΔP, with viscoelastic behavior (here described by the viscosity μ). When the heating process is mainly carried out using high-pressure saturated steam (in the case of DIC), the increase in temperature is closely related to an increase in the moisture content of the material:

Increase in water content during saturated steam heating:

$$\Delta W_{\mathrm{cond}} = \frac{m_{\mathrm{v}}}{m_{\mathrm{s}}} = \frac{\left(c_{\mathrm{p,s}} + W_{\mathrm{o}} c_{\mathrm{p,w}}\right)(T_{\mathrm{t}} - T_{\mathrm{i}})}{L} > 0 \tag{23}$$

where m_{v} is the mass of steam condensed on the surface of the product capable of ensuring the change from the initial temperature T_{i} to the treatment temperature T_{t}; m_{s} the mass of dry matter in the treated product; W_{o} the initial moisture of the product to be treated in % of dry matter.

The moisture content in the material just before the pressure drops:

$$W_{\mathrm{i}} = W_{\mathrm{o}} + \Delta W_{\mathrm{condensation}} \tag{24}$$

Glass Transition

High levels of humidity and temperature just before the instant pressure drop often increase the possibility that the material exhibits viscoelastic behavior. Preservation of the new expanded honeycomb structure is closely related to the glass transition of the material; the instant pressure drop, which is responsible for the decrease in both the humidity and temperature of the material, can often increase the likelihood of glass transition. Indeed, experimental measurements of the glass transition temperature (Fig. 1) obtained at each moisture content can be correlated with the Gordon–Taylor relationship (Gordon and Taylor 1952), formulated according to Allaf (1982):

Glass transition: Gordon–Taylor correlation model:

$$T_g(W) = \frac{T_{g,m} + kWT_{g,w}}{1 + kW} \quad (25)$$

where W is the moisture content (dry basis); $T_{g,m}$ and $T_{g,w}$ the glass transition temperatures of the dry material and pure water, respectively. Note that $T_{g,w}$ is suggested by Orford et al. (1990) to be: $T_{g,w} = -139\ ^\circ C$. Finally, k is the Gordon–Taylor correlation parameter, depending on the material. For $k = 1$, the glass transition temperature can be obtained as follows:

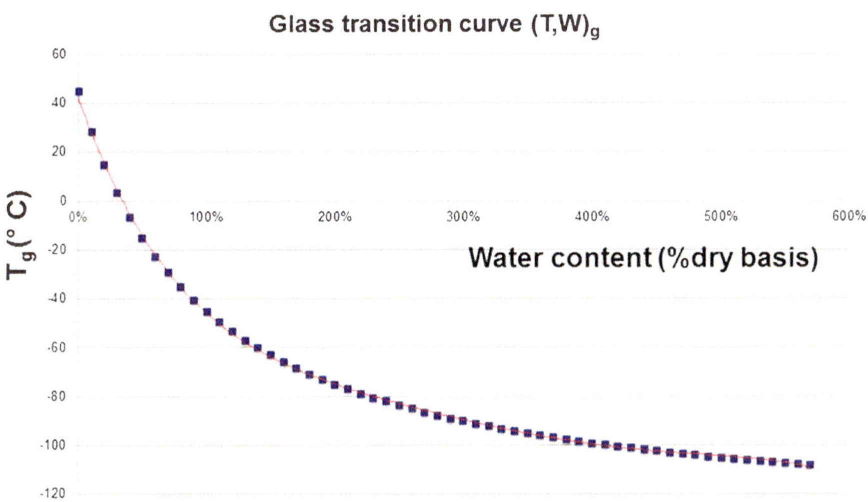

Fig. 1 Change in glass transition vs. water content dry basis W

2.4 Examples of Intensification

DIC treatment can expand the structure of a plant at various levels that can be controlled and are highly dependent on the operating parameters. Since such structural modifications can vary from the simple expansion of the interstitial zone to breaking cell walls, it is possible to control various functional characteristics and technological aptitudes. Direct correlations should be established between:

- Plant anatomy data such as the localization of active molecules, cell forms, size and distribution, and the mechanics and permeability of the cell walls (Amor et al. 2008)

- Structural features such as absolute expansion ratio and porosity, specific surface area, and pore-size distribution
- Functional characteristics such as starting accessibility, diffusivity and permeability within the matrix, and solute yields and availability of the desired compounds

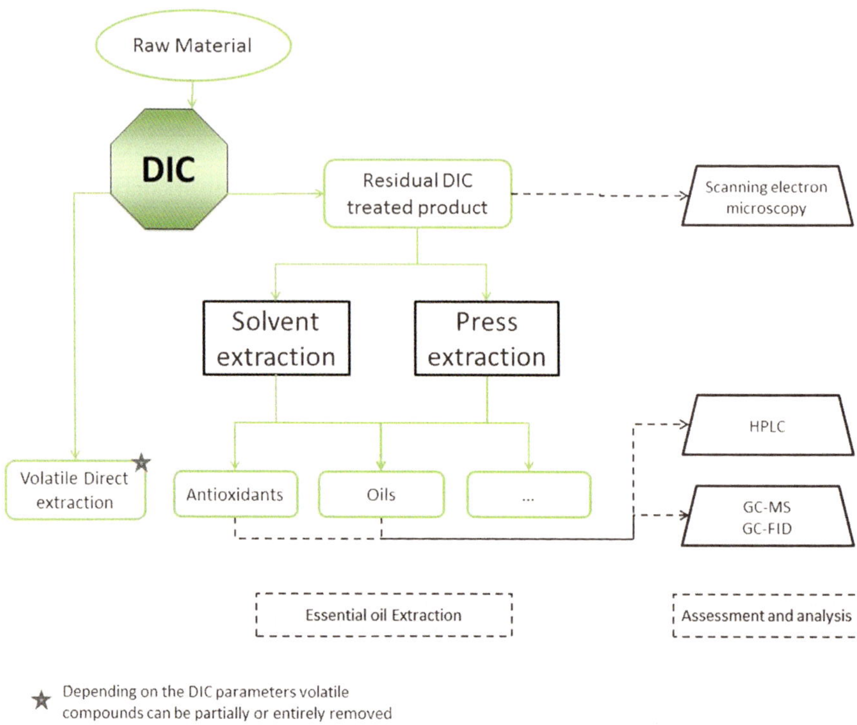

Fig. 2 General treatment including DIC-assisted solvent extraction

In all solvent extraction operations with a plant substance, DIC texturing plays an important role as an intensification step (Fig. 2). It should be studied and optimized depending on the raw material, unit operation performances, and final solute/product quality.

2.4.1 Vegetable Seed Oil: Canola Seed

The starting accessibility δX_s and effective diffusivity D_{eff} values were measured with a view to canola oil extraction with hexane. DIC could increase the 6 h yields by 150 % (Table 1), without modifying the fatty acid profile.

Table 1 Starting accessibility, effective diffusivity, and oil yield

		DIC	RM
Starting accessibility δX_s	Value (g/100 g dm)	15	10
	Ratio (%)	29	19
	Enhancement	150 %	100 %
Effective diffusivity D_{eff}	D_{eff} (10^{-12} m^2 s^{-1})	1.9	0.7
	Enhancement	266 %	100 %
6-h oil yield X_∞	$Yield_{oil}$ (g/100 g dm)	37	25
	Enhancement	150 %	100 %

RM raw material, *DIC* seeds treated by DIC

DIC texturing treatment significantly improved the starting accessibility, with more than 29 % of canola oil—that is, up to 15 g oil/100 g dm—extracted during the short first extraction stage, compared with 19 % for untreated seeds, or 10 g oil/100 g dm.

For both sets of samples, the remaining oil was extracted via a diffusion procedure.

Finally, it is worth pointing out the direct impact of optimized DIC treatment on reducing the solvent/material concentration ratio, which greatly reduces the energy consumption of the total operation.

Since DIC is a high-temperature short-time process with an abrupt drop in cooling pressure towards a vacuum, optimized DIC treatment avoids any discernable thermal degradation (Table 2).

Table 2 Fatty acid profile (relative %) obtained via gas chromatography

%	RM	DIC
C16	4.16	4.32
C16:1n9	0.18	0.19
C18	1.44	1.43
C18:1n9	57.99	57.58
C18:2n6	21.89	21.79
C18:3n3	9.19	9.45
C20	0.53	0.51
C20:1n9	2.02	2.10
C20:2n6	0.12	0.13
C22	0.29	0.25
C22:1n9	2.10	2.16
Saturated fatty acid (SFA)	6.42	6.51
Mono-unsaturated fatty acid (MUFA)	62.29	62.03
Poly-unsaturated fatty acid (PUFA)	31.21	31.37

2.4.2 Extraction of Active Molecules

Solvent extraction of active molecules from plants was studied. DIC treated and untreated plant materials were compared in terms of flavonoid and anthocyanin extraction (Figs. 3, 4, and 5). Systematically, appropriate DIC texturing noticeably improved the extraction kinetics of both anthocyanins and flavonoids, while reducing extraction time (Table 3). DIC treatment significantly increased the antioxidant activity (AOA) of myrtle and buckthorn (3.57 and 68.62 times higher than BHT, respectively).

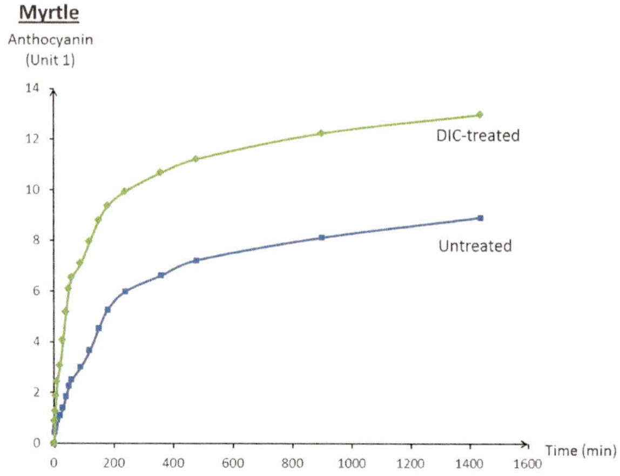

Fig. 3 Solvent extraction of anthocyanins from DIC treated and untreated myrtle leaves (unit 1: mg eq. kaempferol/g dm)

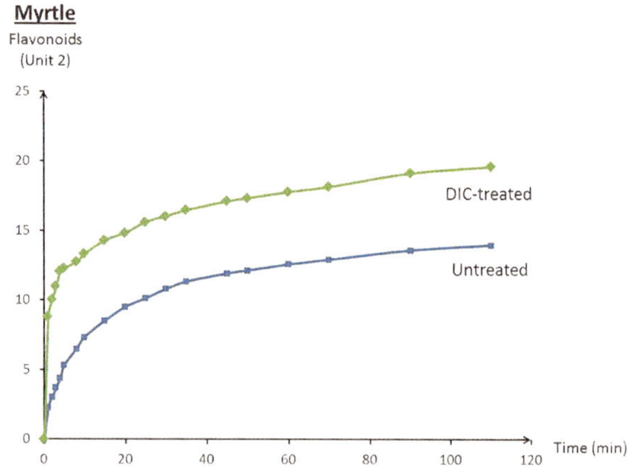

Fig. 4 Solvent extraction of flavonoids from myrtle leaves. DIC treated and untreated samples (unit 2: mg eq. kaempferol/g dm)

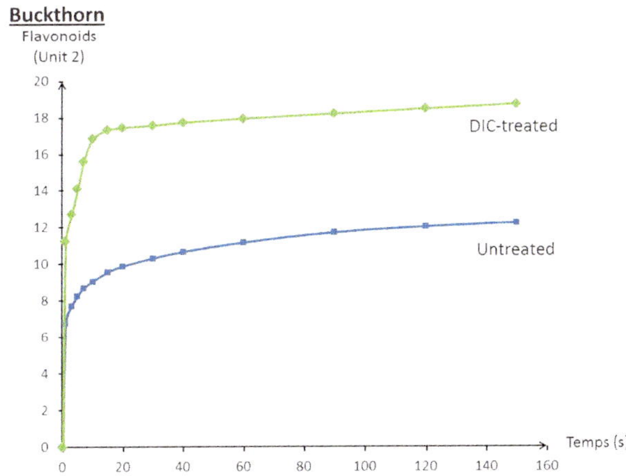

Fig. 5 Solvent extraction of flavonoids from buckthorn. DIC treated and untreated samples (unit 2: mg eq. kaempferol/g dm)

Table 3 Extraction kinetics: Extraction time (min), and percentage enhancement of starting accessibility δX_s and effective diffusivity D_{eff} and anti-oxidant activity (AOA) using DIC to treat myrtle and buckthorn leaves compared with untreated leaves

	Myrtle leaves (*Myrtus*)		Buckthorn (*Rhamnus alaternus*)
Extraction kinetics	Flavonoids[a]	Anthocyanins[b]	Flavonoids[a]
$\%\delta X_s$	344	277	165
$\%D_{eff}$	286	344	471
Time (min)	14 instead of 112	160 instead of 1,450	2 instead of 150
AOA (μg mL^{-1})	0.78 ± 0.05 instead of 2.18 ± 0.07		0.12 ± 0.06 instead of 8.22 ± 0.01
	357 % higher than BHT[c]		6,862 % higher than BHT[c]

[a]Methanol–water (80:20, v/v) and 1.6 % HCl at ambient temperature
[b]Methanol:0.1 % HCl (volume:volume) at ambient temperature
[c]Amount of BHT for %IC50 level of antioxidant effect is 48.13 ± 0.13 μg mL^{-1}

3 Deodorization Before Extraction of Bioactive Molecules

In some cases, the antioxidant substances must be potent at low quantities given the fact that changes in foodstuff aromas must be kept to a minimum (Pokorny et al. 2001). Indeed, an antioxidant ingredient that still has an aroma will trigger taste modification when added to foodstuffs (Ares et al. 2009). As a consequence, processes for obtaining such antioxidants should reduce or even completely remove aromas, while preserving high AOA.

In terms of aroma extraction, the deodorization method often used in industry is hydrodistillation, although it is time-consuming and involves issues such as thermal degradation of antioxidant molecules and loss of yield.

In contrast, DIC autovaporization is a quick and highly effective deodorization, which can simultaneously modify the plant structure at various levels that can be controlled and are highly dependent on the operating parameters (Ben Amor and Allaf 2009; Amor et al. 2008).

Such modifications, from the interstitial zone to broken cell walls, result in severe modifications of the structural characteristics. It should then be possible to control the functional features and technological capacities that help to greatly intensify mass transfer by improving starting accessibility, diffusivity, and permeability within the matrix.

Whether with rosemary leaves or orange peel, SEM images illustrate these profound structural differences between the raw material and samples treated using hydrodistillation or DIC, with an obvious direct impact in terms of active molecule extraction.

3.1 Rosemary Leaves

Rosemary (*Rosmarinus officinalis*) is known to have a strong AOA (Erkan et al. 2008). The main substances responsible for this activity are phenolic diterpenes (Schwarz and Ternes 1992a, b; Schwarz et al. 1992) such as carnosol, rosmanol, carnosic acid, methyl carnosate, and phenolic acids such as rosmarinic acid (Petersen and Simmonds 2003). However, rosemary leaves contain between 0.7 and 3 % fresh weight of essential oils, depending on the variety, the harvesting method, etc. (Bousbia et al. 2009; Sui et al. 2012; Singh and Guleria 2013; Szumny et al. 2010).

After the extraction of essential oils by DIC or by hydrodistillation, SEM images showed profound structural differences compared with the raw material (Fig. 6). Before DIC treatment, the product had a slightly compact and relatively well-organized cell structure. The impact of DIC treatment is clear from the evident expansion of the product. On the other hand, hydrodistillation results in visible shrinkage, which reduces diffusivity and hence should slow down the extraction of antioxidants after deodorization.

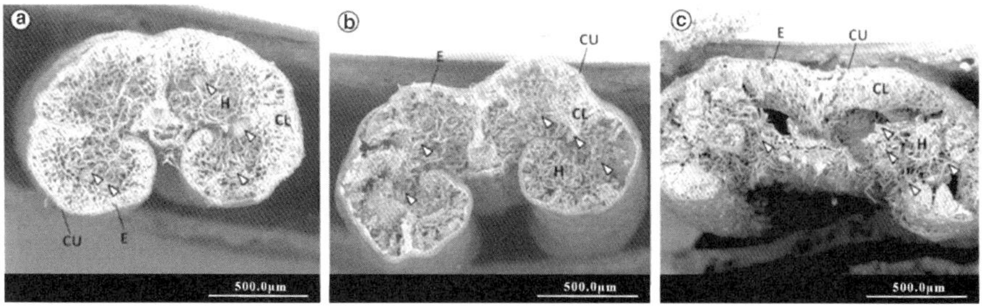

Fig. 6 Scanning electron micrographs of rosemary leaves: (**a**) untreated; (**b**) hydrodistillated; and (**c**) DIC treated. *CU* cuticle, *E* epidermis, *H* hairs, *CL* collenchyma, ▷ glandular trichomes

Antioxidant ethanol:water (80:20) ambient temperature extraction was carried out on rosemary leaves treated using hydrodistillation or DIC deodorization, or on raw material. Extracts were then analyzed and identified by HPLC (Table 4).

Table 4 Comparative study of rosmarinic acid extraction after HD or DIC as deodorization technology

Deodorization technology	Rosmarinic acid (mg/g dm)	Deodorization treatment time
Untreated	6.74 ± 0.20	–
HD	1.92 ± 0.29	4 h
DIC	12.76 ± 0.22	3 min

3.2 *Orange Peel*

Citrus is an important crop, with over 100 million tons produced worldwide every year (FAO 2003, 2010; USDA 2011;). There are large amounts of by-products such as peel and seed representing about half of the fruit weight (Manthey and Grohmann 2001). Processing citrus by-products represents a rich and growing source of natural target molecules, such as phenolic compounds (Rouseff et al. 1987). The outer peel of citrus, known as flavedo, has a large number of very small essential oil glands. The conventional techniques used to extract these essential oils are mechanical, or use hydro- or steam-distillation, which require grinding because of the structural resistance that hinders the easy extraction of essential oil (Atti-Santos et al. 2005).

The natural structure of citrus peel hinders the easy extraction of essential oil. Extraction using hydrodistillation showed that grinding was a necessary pretreatment, but took 4 h. In contrast DIC, carried out on pieces of orange peel, acts as a thermomechanical treatment. DIC enables the extraction of essential oils in less than 3 min. SEM showed a slightly compact and relatively well-organized cell structure of the raw material (Fig. 7). DIC led to an evident expansion of the well-organized structure. On the other hand, hydrodistillation shrinkage is clearly noticeable, and this weakened diffusivity and hence resulted in poor extraction kinetics for both essential oils (deodorization step) and antioxidants (solvent extraction after the deodorization step).

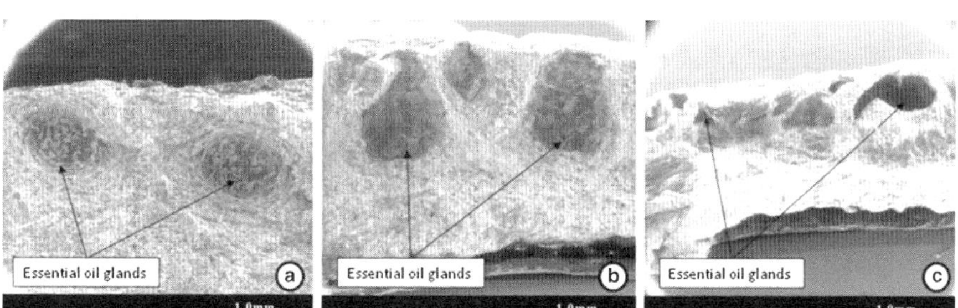

Fig. 7 Micrographs of dried orange peel: (**a**) untreated raw material; (**b**) DIC treated peel; (**c**) hydrodistillated peel

After a "deodorization" step, the material could be recovered and the solvent extraction of the active molecules (such as antioxidants) had higher kinetics than the raw material or the hydrodistillated material (Table 5).

Table 5 Orange peel naringin hesperidin extraction kinetics after 1-h solvent extraction

Deodorization technology	Naringin (g/100 g dm)	Hesperidin (g/100 g dm)	Deodorization treatment time
Untreated	3.3 ± 0.3	34 ± 3	–
HD	1.2 ± 0.1	10.2 ± 0.6	4 h
DIC	5.7 ± 0.2	64 ± 3	2 min

4 Closing Remarks

A review of the scientific literature regarding DIC-assisted solvent extraction from plants shows that specific fundamental and process engineering research was carried out which defined appropriate intensification processes for both effective deodorization and short-time/high-yield active extraction processes. This led to the construction of various laboratory scale and high-capacity industrial reactors.

Whatever the botanical variety and the plant organ, the natural structure of plants is too compact. DIC systematically provided adequate texturing of the matrix to enable a greater expansion, resulting in a better technological aptitude thanks to a higher diffusivity, starting accessibility, and availability of numerous active molecules.

Specific kinetics studies suggested that exceptionally appropriate extraction processes could be defined by coupling DIC texturing with new technologies, e.g., ultrasound and supercritical fluid. DIC transforms these technologies into more effective and easily manageable solutions for large-scale industrial production technologies. These coupled technologies have increased the understanding of numerous academic research results and contributed to a highly relevant technology transfer to industrial manufacturing. Furthermore, regarding the botanical and structural sides, the specificity of DIC as a highly controlled instantaneous texturing and autovaporization process means that the natural structure can be finely modified to perfectly adapt it for the extraction of active molecules from plants. The transfer to industry is then clearly defined and implemented.

Armed with the fundamental analyses, experimental results, empirical models, and botanical structure, academic researchers and industrial engineers have perfectly defined different extractions from plants. This research work should be extended, especially to supercritical fluid extraction. However there are already very good approaches that also take into account the microscale botanical aspects.

This perfectly controlled DIC technology, used as a simple process or coupled to other new technologies, could be successfully transferred for solvent extractions from plants.

Different standard apparatus have been designed, built, and proposed in combination with appropriate experiments and a rigorous design to improve process performances and the quality of the end product.

References

Aguilera JM, Stanley DW (1999) Microstructural principles of food processing and engineering. Aspen Publication, USA

Allaf K (1982) Transfer phenomena and industrial applications. Lebanese University Faculty of Science, Beirut

Allaf K (1988) Approche à l'analyse fondamentale de l'expansion par alvéolation selon différents procédés (puffing, cuisson-extrusion...). Université de Technologie de Compiègne, Compiègne, France

Allaf K (2002) Analysis of instantaneity in thermodynamic processes; Fundamental laws. Paper presented at the first Franco-Lebanese symposium on technologies and studies on process engineering and biochemistry, Beyrouth, 22–25 July 2002

Allaf K, Vidal P (1989) Feasibility study of a new process of drying/swelling by instantaneous decompression towards vacuum. Chemical Engineering Department, University of Technology of Compiègne, Compiègne, France

Allaf K, Louka N, Bouvier JM, Parent F, Forget M (1993) Procédé de traitement de produits biologiques en vue de la modification de leur texture, installations pour la mise en oeuvre d'un tel procédé et produits ainsi réalisés. 93/09728

Allaf K, Debs-Louka E, Louka N, Abraham G (1998) Procédé de réduction ou d'élimination d'organismes, de microorganismes, de pasteurisation et de stérilisation des produits solides en morceaux ou pulvérulents et installation pour la mise en œuvre d'un tel procédé. 98/02032

Allaf K, Besombes C, Berka-Zougali B, Kristiawan M, Sobolik V, Allaf T (2011) Instant controlled pressure drop technology in plant extraction processes. In: Lebovka N, Vorobiev E, Chemat F (eds) Enhancing extraction processes in the food industry, Contemporary food engineering series. CRC Press/Taylor & Francis Group, Dublin, Ireland, pp 255–302

Allaf T, Mounir S, Tomao V, Chemat F (2012) Instant controlled pressure drop combined to ultrasounds as innovative extraction process combination: fundamental aspects. Procedia Eng 42:1164–1181. doi:10.1016/j.proeng.2012.07.498

Amor BB, Lamy C, Andre P, Allaf K (2008) Effect of instant controlled pressure drop treatments on the oligosaccharides extractability and microstructure of Tephrosia purpurea seeds. J Chromatogr A 1213(2):118–124. doi:10.1016/j.chroma.2008.10.065

Ares G, Barreiro C, Deliza R, Gámbaro A (2009) Alternatives to reduce the bitterness, astringency and characteristic flavour of antioxidant extracts. Food Res Int 42(7):871–878. doi:10.1016/j.foodres.2009.03.006

Arhaliass A, Legrand J, Vauchel P, Fodil-Pacha F, Lamer T, Bouvier JM (2009) The effect of wheat and maize flours properties on the expansion mechanism during extrusion cooking. Food and Bioprocess Technology 2(2):186–193. doi:10.1007/s11947-007-0038-6.

Atti-Santos AC, Rossato M, Serafini LA, Cassel E, Moyna P (2005) Extraction of essential oils from lime (Citrus latifolia Tanaka) by hydrodistillation and supercritical carbon dioxide. Braz Arch Biol Technol 48(1):155–160

Ben Amor B (2008) Maîtrise de l'aptitude technologique de la matière végétale dans les opérations d'extraction de principes actifs; texturation par Détente Instantanée Contrôlée DIC. Université de La Rochelle, La Rochelle, France

Ben Amor B, Allaf K (2009) Impact of texturing using instant pressure drop treatment prior to solvent extraction of anthocyanins from Malaysian Roselle (*Hibiscus sabdariffa*). Food Chem 115(3):820–825. doi:10.1016/j.foodchem.2008.12.094

Bousbia N, Abert Vian M, Ferhat MA, Petitcolas E, Meklati BY, Chemat F (2009) Comparison of two isolation methods for essential oil from rosemary leaves: hydrodistillation and microwave hydrodiffusion and gravity. Food Chem 114(1):355–362. doi:10.1016/j.foodchem.2008.09.106

Crank J (1975) The mathematics of diffusion

Debs-Louka E (2000) Destruction des microorganismes par voie thermo-mécanique contrôlée dans des produits solides en morceaux ou en poudre. Application aux épices et aromates. Université de La Rochelle, La Rochelle, France

Delgado-Rosas M (2002) Analyse et conception d'un réacteur à pulvérisation contrôlée sous vide (VCS). Université de La Rochelle, La Rochelle, France

Erkan N, Ayranci G, Ayranci E (2008) Antioxidant activities of rosemary (*Rosmarinus officinalis* L.) extract, blackseed (*Nigella sativa* L.) essential oil, carnosic acid, rosmarinic acid and sesamol. Food Chem 110(1):76–82. doi:10.1016/j.foodchem.2008.01.058

FAO (2003) Medium-term prospects for agricultural commodities. Food and agriculture organization of the united nations, US

FAO (2010) Citrus Fruit: Fresh and Processed - Annual Statistics. Food and agriculture organization of the united nations, US

Fernández MB, Perez EE, Crapiste GH, Nolasco SM (2012) Kinetic study of canola oil and tocopherol extraction: parameter comparison of nonlinear models. J Food Eng 111 (4):682–689. doi:10.1016/j.jfoodeng.2012.01.036

Gordon M, Taylor JS (1952) Ideal copolymers and the second-order transitions of synthetic rubbers. i. non-crystalline copolymers. Journal of Applied Chemistry 2(9):493–500. doi:10.1002/jctb.5010020901.

Habba A (1997) Nouveaux procédés de précuisson et d'étuvage du riz par détente instantanée contrôlée (DIC): analyse des opérations, optimisation des procédés et approche à l'industrialisation. Université de Technologie de Compiègne, Compiègne, France

Haddad J (2002) Impacts des procédés hydro-thermo-mécaniques dans la valorisation des graines oléoprotéagineuses. Université de La Rochelle, La Rochelle, France

Iguedjtal T, Louka N, Allaf K (2008) Sorption isotherms of potato slices dried and texturized by controlled sudden decompression. J Food Eng 85(2):180–190. doi:10.1016/j.jfoodeng.2007.06.028

Juhel F (2000) Etude de l'application de deux nouvelles opérations de séchage/texturation du poisson; analyse comparative du procédé et du produit. Université de La Rochelle, La Rochelle, France

Leybros J, Frémeaux P (1990) Extraction solide–liquide. Aspects théoriques. Techniques de l'Ingénieur J1 077 06

Louka N, Allaf K (2004) Expansion ratio and color improvement of dried vegetables texturized by a new process "controlled sudden decompression to the vacuum": application to potatoes, carrots and onions. J Food Eng 65(2):233–243. doi:10.1016/j.jfoodeng.2004.01.020

Manthey JA, Grohmann K (2001) Phenols in citrus peel byproducts. Concentrations of hydroxycinnamates and polymethoxylated flavones in citrus peel molasses. J Agric Food Chem 49(7):3268–3273

Mounir S (2007) Studies of new manufacturing process of powders by inserting the instant controlled pressure drop DIC within spray-drying, swell-drying and controlled vacuum atomization of dairy products. Université de La Rochelle, La Rochelle, France

Mounir S, Allaf K (2008) Three-stage spray drying: new process involving instant controlled pressure drop. Drying Technol 26(4):452–463. doi:10.1080/07373930801929334

Mounir S, Allaf K (2009) Study and modelling of dehydration and rehydration kinetics within porous medium. In: Proceedings of European Drying Conference AFSIA, Lyon, France

Orford PD, Parker R, Ring SG (1990) Aspects of the glass transition behaviour of mixtures of carbohydrates of low molecular weight. Carbohydrate Research 196 (0):11–18. doi:10.1016/0008-6215(90)84102-z

Petersen M, Simmonds MSJ (2003) Rosmarinic acid. Phytochemistry 62(2):121–125. doi:10.1016/s0031-9422(02)00513-7

Pokorny J, Yanishlieva N, Gordon M (2001) Antioxidants in Foods: Practical Applications. Woodhead Publishing in Food Science and Technology. Woodhead Publishing Ltd, England

Rakotozafy H (2001) Application du nouveau procédé de déshydratation par détentes successives (DDS), dans le séchage de produits biologiques à haute valeur ajoutée. Université de La Rochelle, La Rochelle - France

Rouseff RL, Martin SF, Youtsey CO (1987) Quantitative survey of narirutin, naringin, hesperidin, and neohesperidin in citrus. J Agric Food Chem 35(6):1027–1030

Sahyoun W (1996) Maîtrise de l'aptitude de matériaux agro-alimentaires aux procédés de séchage. Etude de l'adéquation entre les états structuraux, biochimiques, physiques et comportementaux sur les processus de déshydratation = Mastering the aptitude of food products to drying processes. Study of the adequacy between structural, biochemical, physical, and behavioural states on the dehydratation process. Université de Technologie de Compiègne, France

Sanya E (2000) Analyse du traitement hydro-amido-thermique de bois gorgé d'eau; application à la restauration-préservation du bois archéologique. Université de La Rochelle, La Rochelle, France

Schwarz K, Ternes W (1992a) Antioxidative constituents of *Rosmarinus officinalis* and *Salvia officinalis*. I. Determination of phenolic diterpenes with antioxidative activity amongst tocochromanols using HPLC. Z Lebensm Unters Forsch 195(2):95–98

Schwarz K, Ternes W (1992b) Antioxidative constituents of *Rosmarinus officinalis* and *Salvia officinalis*. II. Isolation of carnosic acid and formation of other phenolic diterpenes. Z Lebensm Unters Forsch 195(2):99–103

Schwarz K, Ternes W, Schmauderer E (1992) Antioxidative constituents of *Rosmarinus officinalis* and *Salvia officinalis*. III. Stability of phenolic diterpenes of rosemary extracts under thermal stress as required for technological processes. Z Lebensm Unters Forsch 195(2):104–107

Schwartzberg HG, Chao RY (1982) Solute diffusivities in leaching processes. Food Technolgy 2:73–86.

Singh M, Guleria N (2013) Influence of harvesting stage and inorganic and organic fertilizers on yield and oil composition of rosemary (*Rosmarinus officinalis* L.) in a semi-arid tropical climate. Ind Crops Prod 42:37–40. doi:10.1016/j.indcrop.2012.04.054

Sui X, Liu T, Ma C, Yang L, Zu Y, Zhang L, Wang H (2012) Microwave irradiation to pretreat rosemary (*Rosmarinus officinalis* L.) for maintaining antioxidant content during storage and to extract essential oil simultaneously. Food Chem 131(4):1399–1405. doi:10.1016/j.foodchem.2011.10.007

Szumny A, Figiel A, Gutiérrez-Ortíz A, Carbonell-Barrachina ÁA (2010) Composition of rosemary essential oil (*Rosmarinus officinalis*) as affected by drying method. J Food Eng 97 (2):253–260. doi:10.1016/j.jfoodeng.2009.10.019

Thai-Cong D (2003) Etude de l'application du procédé hydrothermique dans le traitement de différents types de riz : procédé d'étuvage et micro- expansion par détente instantanée contrôlée et impact sur les propriétés fonctionnelles. Université de La Rochelle, La Rochelle - France

USDA (2011) Citrus: world markets and trade. United States Department of Agriculture—Foreign Agricultural Service, Washington, DC

Coupling DIC and Ultrasound in Solvent Extraction Processes

Tamara Allaf, Colette Besombes, Valérie Tomao, Farid Chemat, and Karim Allaf

1 Introduction

Intensification of the main industrial unit operations is designed to reduce treatment time, decrease energy consumption, and preserve or even improve the multidimensional quality of the end products. The philosophy behind intensification involves identifying different processes and determining their possible interaction. In order to do so, it is first essential to shed light on the limiting phenomena in order to intensify them, thereby improving the kinetics of the entire operation.

2 Fundamental Aspects

2.1 Solvent Extraction

Solvent extraction involves a solid/liquid interaction that removes the soluble components from solids. This operation, from a technological point of view, is a phenomenon that involves the diffusion of a carrier fluid (liquid) through a porous solid. The carrier fluid transports a solvent capable of "dissolving" one or more specific molecules from a solid (plant) or a liquid. The solute-in-solvent is also transferred within the porous solid through a diffusion process.

T. Allaf (✉)
ABCAR-DIC Process, La Rochelle, France
e-mail: tamara.allaf@gmail.com

C. Besombes • K. Allaf
Laboratory of Engineering Science for Environment (LaSIE FRE 3474) CNRS, 17042 La Rochelle Cedex 01, University of La Rochelle, La Rochelle, France
e-mail: kallaf@univ-lr.fr

V. Tomao • F. Chemat
Université d'Avignon et des Pays de Vaucluse, INRA, UMR408, 84000 Avignon, France

T. Allaf and K. Allaf (eds.), *Instant Controlled Pressure Drop (D.I.C.) in Food Processing: From Fundamental to Industrial Applications*, Food Engineering Series, DOI 10.1007/978-1-4614-8669-5_8,

The gradient of solute concentration between the solution at the surface in the solid phase (more concentrated) and the liquid phase results in a transfer by diffusion of these dissolved active molecules to the surroundings. At the end of the operation, the system tends toward equilibrium and the diffusion transfer becomes nil.

In contrast, if the liquid phase is continuously renewed, the diffusion continues until the complete exhaustion of the solid phase. At the end of the operation, the solid (residue), inert or insoluble, contains very little or no solute at all (Allaf et al. 2011).

Part III—Chap. 3—presents solvent extraction transfer in detail so we will only highlight the general elements of this process in this chapter.

A first stage of solute dissolution in the solvent is carried out at the surface of the product (illustrated by the starting accessibility). It is followed by diffusion phenomena of the solvent towards the core of the solid matrix and of the solute within the filled-with-solvent pores. This specific transport occurs as a Fick-type diffusion process.

The type of solvent used and its polarity depend on the solute that is to be extracted, and the temperature is limited by the boiling temperature and the thermal sensitivity of the compounds. Thus, it can be assumed that the interaction between the solvent and the product results in the solute directly reaching equilibrium.

Agitation of the solvent in the external environment allows the solute that is accessible at the exchange surface to be easily and quickly extracted and transported far from the exchange surface. This first intensification of the solvent extraction process significantly replaces natural convection. Internal solvent diffusion, followed by solute-in-solvent diffusion within the matrix, becomes the limiting processes.

In this chapter, we will therefore be focusing on the intensification of this specific limiting process.

2.2 Extraction Intensification

Subsequently, solvent extraction can be intensified in several ways:

A *grinding process* that increases the exchange surface, thus augmenting starting accessibility while maintaining effective diffusivity constant	*DIC* texturing to expand the granules and increase the material's porosity. Thus, the effective diffusivity of both solvent and solute-in-solvent within the plant medium also increases. Indeed, the natural structure of vegetables and more specifically the cytoplasmic membrane and the cellular wall cannot support the liquid transfer processes. The resistance of the structure often seems to be the principal restricting factor for the kinetics of the operation

(continued)

(continued)

Solvent agitation to allow more effective transport of solute from the product surface towards the external medium by convection instead of diffusion *Renewing the solvent* at low solute concentration by a counter-current exchange system, etc.	*Ultrasound* (US) to establish an internal micro-convection within the pores instead of the very prejudicial diffusion process

3 Ultrasound: Fundamental Aspects

3.1 Principle

Ultrasounds are mechanical waves that require an elastic medium to spread. Ultrasound frequencies are above the upper limit of human hearing (from 16 Hz to 16–20 kHz) and below the frequency of microwaves (20 kHz to 10 MHz). Ultrasounds can be divided into two groups with regard to their frequencies:

- Diagnostic ultrasounds: these ultrasounds are also called low-power or high-frequency ultrasounds. They range from 5 to 10 MHz and are used in several fields such as medical imaging (med scan) or even for defect detection (bond inspection for plastics). Diagnostic ultrasounds are used to make measurements without altering the medium, unlike power ultrasounds.
- Power ultrasounds or low-frequency ultrasounds are used to produce physical or chemical effects in the medium. They are used in sonochemistry (to facilitate or accelerate chemical reactions), agriculture (water dispersion), and in industry (cutting, plastic welding).

In order to visualize an ultrasonic effect, its action can be described as a piston on the surface of the medium (Pingret et al. 2013) (Fig. 1). This results in a succession of compression and rarefaction phases in the medium. When the piston is in the open position it induces a compression in the medium and when the piston is in the contracted (pull) position it creates a rarefaction phase.

On a molecular scale, as the sound wave passes through the medium, the molecules are temporarily dislodged from their original position and can enter into collision with the surrounding molecules (Vinatoru 2001) (Fig. 1). Then, during the rarefaction phase the first group of molecules will be pulled back towards their original position and the kinetic energy involved will pull them further back from this position. This will create regions of rarefaction in the medium but due to their elastic properties all the molecules will go back to their original location when the sound wave has fully passed through the medium.

With regard to solvent extraction, oscillatory particle motion produced by high-intensity ultrasonic waves can also induce secondary flows, known as acoustic streaming. This kind of agitation produces an internal convection motion of the solute within the solvent which is inside the holes of the porous material. Moreover, cavitation produces micro-jets at the surface of the food material that may increase the exchange surface, the effect of which has to be revealed through a high starting accessibility. Both effects can increase mass transfer of solute within the solvent which is present within the solid (Mason 2000; Toma et al. 2001).

In other words, ultrasonic treatment involves an internal motion/agitation of the liquid within the pores, resulting in a solute transfer by convection instead of or coupled with diffusion (Toma et al. 2001).

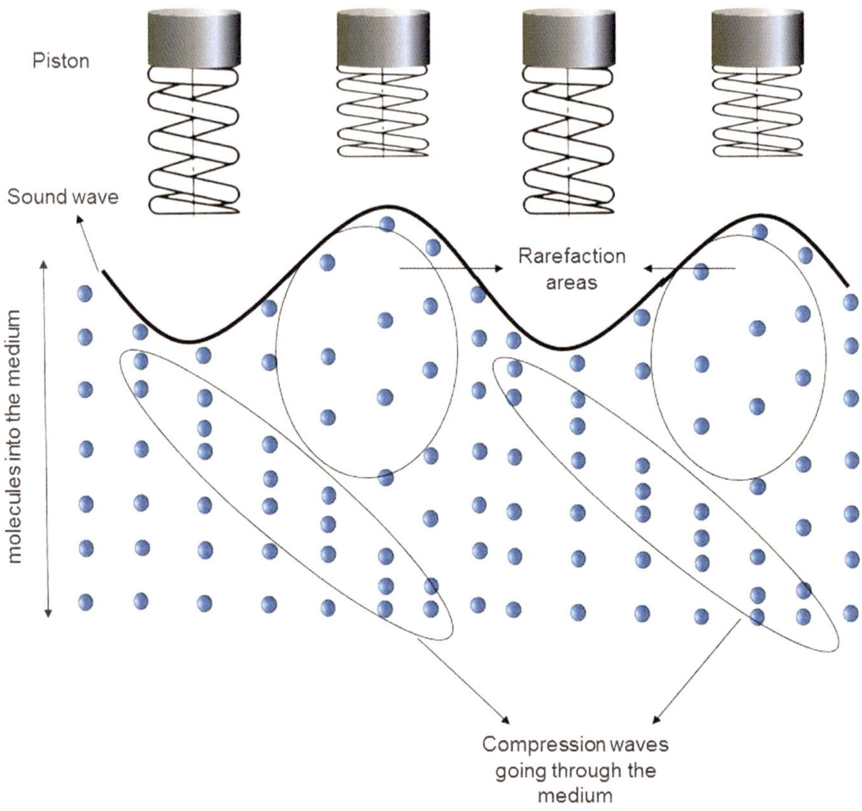

Fig. 1 Adiabatic compression and rarefaction cycles induced by a sound wave

3.2 *Mass Transfer*

3.2.1 External Transport

Transport of a solute in an external solvent from the surface of a solid is possible by *diffusion* from the zone of high concentration of solute (close contact with the solid exchange surface) toward the low concentration zone, with the gradient of solute concentration as the driving force.

$$\frac{\rho_{\text{solute}}}{\rho_{\text{solvent}}}\left(\vec{v}_{\text{solute}} - \vec{v}_{\text{solvent}}\right) = -D_{\text{solute}-\text{solvent}}\vec{\nabla}\left(\frac{\rho_{\text{solute}}}{\rho_{\text{solvent}}}\right) \tag{1}$$

Different methods are generally used to intensify these external transfer phenomena, such as:

- Choosing the best solvent in terms of:
  - Obtaining the highest solute dissolution
  - Obtaining the lowest viscosity to achieve the best diffusivity (D) of solute within the solvent
- Establishing external mechanical agitation to achieve external mass transport by *convection* rather than *diffusion*
- Renewing external solvent to reduce solute concentration:
  - Improving the difference between the exchange surface and the total solvent concentration
  - Maintaining the concentration of solute in the solvent as far as possible from the saturation level

3.2.2 Internal Transfers

Once the external transport is adequately intensified, internal transfers become the limiting processes.

Two main types of internal transfer phenomena have to be analyzed: the solvent to be transferred from the exchange surface to the internal solid matrix and the solute to be transferred within the solvent situated inside the holes of the porous material. The choice of an appropriate intensification method depends on two phenomena:

1. The first point to be noted in an intensification process is that in the transfer of solvent from the exchange surface to the internal solid matrix, the various processes of diffusion, capillarity, and osmosis from a high solvent concentration toward a low solvent concentration occur with the gradient of solvent concentration as the driving force and with D_{eff} as effective diffusivity.

$$\frac{\rho_{\text{solvent}}}{\rho_{\text{solid}}}\left(\vec{v}_{\text{solvent}} - \vec{v}_{\text{solid}}\right) = -D_{\text{eff}}\vec{\nabla}\left(\frac{\rho_{\text{solvent}}}{\rho_{\text{solid}}}\right) \tag{2}$$

   The intensification must then be achieved by:

   (a) Expanding the matrix structure by détente instantanée contrôlée (DIC, which is French for "instant controlled pressure drop") to increase effective diffusivity D_{eff} and the effective exchange surface.

(b) Selecting the lowest viscosity solvent to increase solvent diffusivity within the solid.
(c) Reducing cell wall resistance through cavitation (ultrasound) with a possible destruction of cell walls, mainly by using a more intense DIC or US treatment.

2. The transfer of solute within the solvent situated inside the holes of porous materials is usually achieved by diffusion from high solute concentration zones toward low solute concentration zones with the gradient of solute concentration in the solvent as the driving force and D as the standard solute–solvent diffusivity.

$$\frac{\rho_{\text{solute}}}{\rho_{\text{solvent}}}\left(\vec{v}_{\text{solute}} - \vec{v}_{\text{solvent}}\right) = -D_{\text{solute-solvent}}\,\vec{\nabla}\left(\frac{\rho_{\text{solute}}}{\rho_{\text{solvent}}}\right) \tag{3}$$

The different intensification methods which can be envisaged:

(a) To expand the matrix structure: since the solute–solvent process closely depends on the amount of solvent inside the holes of the porous material, the solvent has to be present in higher concentrations inside the solid and this is achieved by increasing porosity using DIC.
(b) To ensure a solute-in-solvent *micro-convection* inside the holes instead of *natural diffusion*: US results in micro-convection in the internal solvent and the operation can be postulated to be a similar type of diffusion within the porous solid, with an effective diffusivity D_{US} that is generally much higher than standard solute–solvent diffusivity D:

$$\frac{\rho_{\text{solute}}}{\rho_{\text{solid}}}\left(\vec{v}_{\text{solute}} - \vec{v}_{\text{solid}}\right) = -D_{\text{US}}\,\vec{\nabla}\left(\frac{\rho_{\text{solute}}}{\rho_{\text{solid}}}\right) \tag{4}$$

Thus the advantages of ultrasonic extraction are: a reduction in both extraction temperature and time; a decrease in the quantity of solvent; and an improvement in solute extraction yield. Oscillatory particle motion produced by high-intensity/variable frequency ultrasonic waves also induces a secondary flow that is known as acoustic streaming.

3.3 Conclusion: Subsequent Combination of DIC and US

To summarize, the expanded structure of DIC products significantly intensifies the extraction kinetics by improving effective diffusivity in a solid, whatever the solvent; solute diffusion in the solvent inside the pore is generally considered to be the limiting process within the expanded structure.

Coupled to ultrasound, internal transfer of solute present within the pore can likewise be intensified by inducing convection transfer rather than diffusion (Amor et al. 2008; Ben Amor and Allaf 2009).

4 Experiments

The following shows experimental results obtained for the extraction of antioxidants using ultrasound with DIC-treated orange peel (Allaf et al. 2012).

Orange peel is an interesting product for several reasons:

1. Orange peel is a by-product and today there is a considerable emphasis on therecovery, recycling, and upgrading of waste products (Garau et al. 2007). This is particularly valid for food and the food processing industry in which waste, effluents, residues, and by-products can be recovered and often upgraded to produce higher value and useful products (Reddy and Yang 2005).
2. Orange peel is a product that contains essential oils and antioxidants. Usually essential oil extraction is achieved after a grinding step. With DIC we can recover essential oil (without grinding) and enhance solvent extraction/ultrasound-assisted extraction (UAE) (Allaf et al. 2013).

To summarize, essential oil extraction was achieved using DIC technology. The DIC-textured solid residue (Fig. 2) was then recovered to extract antioxidants using both solvent extraction and UAE.

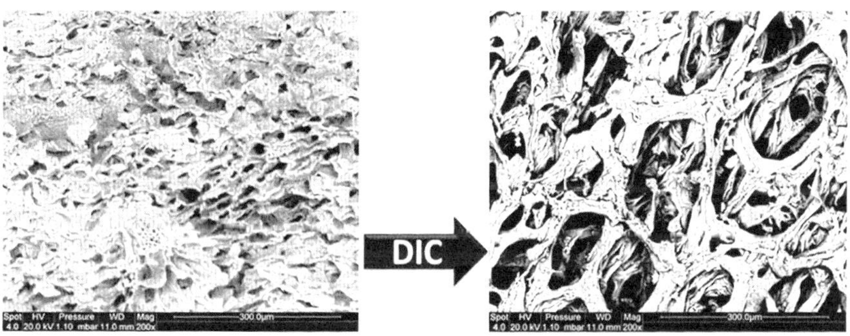

Fig. 2 Scanning electron microscopy of untreated (*left*) and DIC-treated (*right*) orange peel

4.1 Comparative Extraction Yields and Kinetics

Yields and kinetics of naringin and hesperidin extracted from untreated and DIC-treated orange peel followed by 1 h of SE or UAE were compared using HPLC analyses.

The flavanone kinetics results are grouped in Fig. 3. The extraction kinetics of naringin and hesperidin were mostly similar, making it possible to perceive and deduce the action of each process used. It is, however, interesting to note that regarding hesperidin extraction, DIC-SE and RM-UAE showed similar trends, yet combining them gave a complementary effect. Indeed, they act on two different aspects to enhance the extraction. DIC opened the cells, enabling an easy diffusion; ultrasound, through the agitation it provokes, generated an internal convection motion of the solute within the solvent.

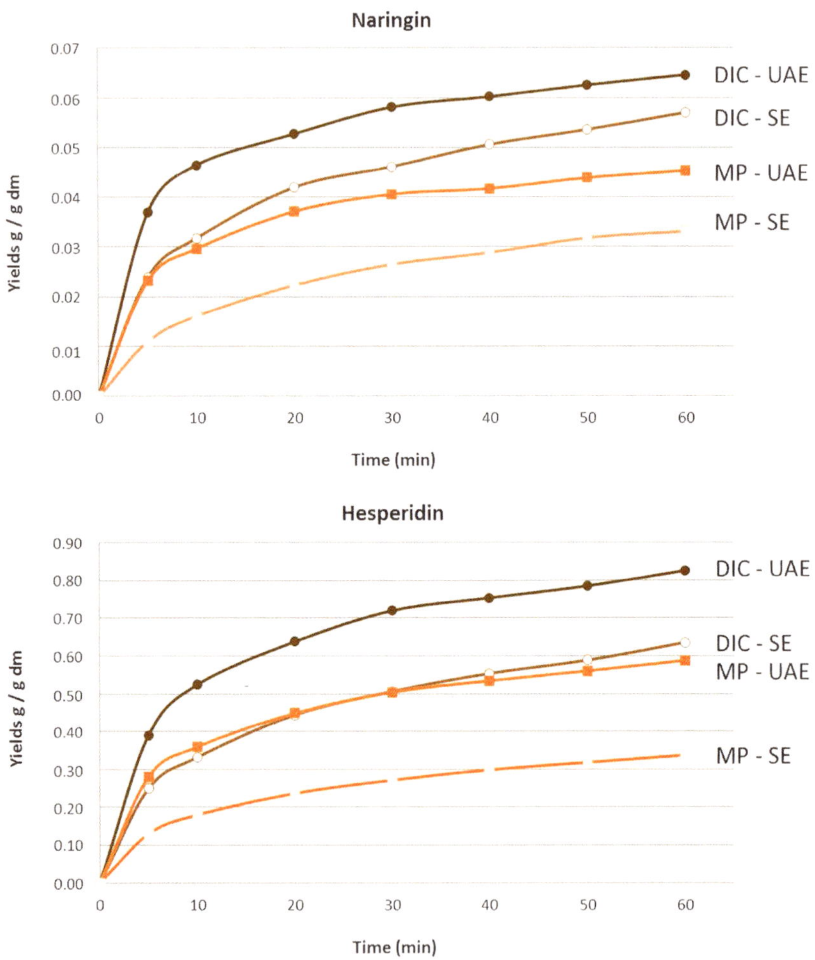

Fig. 3 Naringin and hesperidin extraction kinetics using HPLC analysis

After 60 min of extraction, it was possible to observe large differences in terms of yield and extraction rate.

Consequently, combining DIC with UAE greatly improved the antioxidant extraction yields with markedly improved kinetics compared to standard processes.

To obtain a thorough kinetics analyses, diffusivity and starting accessibility were calculated. The results of a modeling analysis of the extraction kinetics based on the exchange surface and internal diffusion revealed by starting accessibility and effective diffusivity, respectively, are shown in Table 1.

Table 1 Diffusivity and starting accessibility of solvent within different matrices—time for getting 95 % of final extraction

Orange peel	Effective diffusivity (D_{eff}) (10^{-11} m^2/s)	Improvement (%)	Starting accessibility δX_S (g/100 g dry matter)	Starting accessibility ratio (%)	Time (min) for getting 95 % of final extraction ($t_{95\%}$)
Naringin					
RM-SE	4.23	100	1.04	16	310
RM-UAE	5.71	135	2.10	32	211
DIC-SE	13.11	310	2.66	40	100
DIC-UAE	25.22	597	4.31	66	47
Hesperidin					
RM-SE	2.74	100	12.17	14	479
RM-UAE	6.31	231	26.99	32	193
DIC-SE	8.91	326	26.64	32	150
DIC-UAE	26.10	954	50.60	60	49

RM-SE raw material extracted by solvent extraction, *RM-UAE* raw material extracted by ultrasound assisted extraction, *DIC-SE* DIC treated material extracted with solvent, *DIC-UAE* DIC treated material extracted by ultrasound assisted extraction

The raw material extracted with standard SE (RM-SE) was taken as a basis for comparison. Regarding naringin extraction, the effective diffusivity and starting accessibility of DIC-UAE were six and four times higher, respectively, than that of RM-SE. With regard to hesperidin extraction, the effective diffusivity and starting accessibility of DIC-UE were 9.5 and 4 times higher, respectively, than that of RM-SE.

The time taken to achieve 95 % of the final extraction of naringin (assuming that the availability was the same for all the samples) resulted in significant differences between the different materials. Indeed, $t_{95\%}$ was around 5 h for RM-SE, 3 h 30 min for RM-UAE, while it was 1 h 30 min for DIC-SE and less than 1 h for DIC combined with UAE.

4.2 Antioxidant Activity

Phenol antioxidants can reduce reactive oxygen species, including free radicals, at variable rates. Antioxidant activities indicated that the extracts had an anti-radical activity that corresponded to the higher antioxidant concentration. The antioxidant activity of each sample was expressed in terms of IC_{50} (micromolar concentration required to inhibit DPPH radical formation by 50 %) (Mimica-Dukic et al. 2004).

Table 2 IC_{50} (%) as indicator of antioxidant activity of matrices (lower is better)

	DIC + UAE	DIC + SE	RM + UAE	RM + SE
IC_{50} (%)	0.54 ± 0.01	0.62 ± 0.02	1.06 ± 0.03	1.20 ± 0.07

As shown in Table 2, ultrasounds enabled increasing the antioxidant activity of the extract by approximately 13 % (IC_{50} was 1.20 ± 0.07 % for the raw material with SE and 1.06 ± 0.03 % with UAE). Moreover, DIC pretreatment triggered a rise of antioxidant activity twice as high as untreated material regarding both SE and UAE; indeed IC_{50} was 0.62 ± 0.02 % with SE and 0.52 ± 0.01 % with UAE.

5 Conclusion

Fundamental studies on solvent extraction allowed us to define the limiting processes, and thus to propose an expansion by instant controlled pressure drop (DIC), subsequently combined with UAE, as a means of intensification. The operations were coupled to increase the kinetics through the parameters of effective diffusivity D_{eff} and starting accessibility δX_s.

Studies on both the fundamental and experimental aspects regarding various other innovative technologies, DIC/US, DIC/SCF (supercritical fluids), and DIC-MW (microwaves), improved the extraction process in terms of energy consumption, yields, δX_s, D_{eff}, etc. and the end product quality in terms of antioxidant activity.

Pretreatment by DIC and UAE can both improve antioxidant extraction yields and kinetics. The highest yield of hesperidin with the best kinetics was obtained by coupling the treatments. We also note that UAE treatment generated a higher antioxidant activity than standard SE and this was even higher when UAE was performed on a DIC-treated matrix. The impact of DIC on the microstructure provided a reliable explanation for our results. The swelling of the cells enabled a better kinetic extraction in terms of diffusivity and starting accessibility. Accordingly, previous experimental results were highlighted here and the fundamental aspects incorporated into the model.

References

Allaf K, Besombes C, Berka-Zougali B, Kristiawan M, Sobolik V, Allaf T (2011) Instant controlled pressure drop technology in plant extraction processes. In: Lebovka N, Vorobiev E, Chemat F (eds) Enhancing extraction processes in the food industry, Contemporary food engineering series. CRC Press/Taylor & Francis, Dublin, Ireland, pp 255–302

Allaf T, Tomao V, Ruiz K, Chemat F (2013) Instant controlled pressure drop technology and ultrasound assisted extraction for sequential extraction of essential oil and antioxidants. Ultrason Sonochem 20(1):239–246. doi:10.1016/j.ultsonch.2012.05.013

Allaf T, Mounir S, Tomao V, Chemat F (2012) Instant Controlled Pressure Drop Combined to Ultrasounds as Innovative Extraction Process Combination: Fundamental Aspects. Procedia Engineering 42:1164–1181. doi:10.1016/j.proeng.2012.07.498

Amor BB, Lamy C, Andre P, Allaf K (2008) Effect of instant controlled pressure drop treatments on the oligosaccharides extractability and microstructure of Tephrosia purpurea seeds. J Chromatogr A 1213(2):118–124. doi:10.1016/j.chroma.2008.10.065

Ben Amor B, Allaf K (2009) Impact of texturing using instant pressure drop treatment prior to solvent extraction of anthocyanins from Malaysian Roselle (*Hibiscus sabdariffa*). Food Chem 115(3):820–825. doi:10.1016/j.foodchem.2008.12.094

Garau M, Simal S, Rossello C, Femenia A (2007) Effect of air-drying temperature on physico-chemical properties of dietary fibre and antioxidant capacity of orange (Citrus aurantium v. canoneta) by-products. Food Chem 104(3):1014–1024

Mason TJ (2000) Large scale sonochemical processing: aspiration and actuality. Ultrason Sonochem 7(4):145–149. doi:10.1016/s1350-4177(99)00041-3

Mimica-Dukic N, Bozin B, Sokovic M, Simin N (2004) Antimicrobial and antioxidant activities of Melissa officinalis L. (Lamiaceae) essential oil. J Agric Food Chem 52(9):2485–2489

Pingret D, Fabiano-Tixier A-S, Chemat F (2013) Ultrasound-assisted Extraction. In: Rostagno MA, Prado JM (eds) Natural Product Extraction: Principles and Applications. Royal Society of Chemistry, London (UK), pp 89–112. doi:10.1039/9781849737579

Reddy N, Yang Y (2005) Biofibers from agricultural byproducts for industrial applications. Trends Biotechnol 23(1):22–27

Toma M, Vinatoru M, Paniwnyk L, Mason TJ (2001) Investigation of the effects of ultrasound on vegetal tissues during solvent extraction. Ultrason Sonochem 8(2):137–142. doi:10.1016/s1350-4177(00)00033-x

Vinatoru M (2001) An overview of the ultrasonically assisted extraction of bioactive principles from herbs. Ultrasonics Sonochemistry 8(3):303–313. doi:10.1016/s1350-4177(01)00071-2

DIC Intensification of the Mechanical Extraction of Lipids by Pressing

Izaskun Pérez, Carlos Bald, Íñigo Martínez de Marañón, and Karim Allaf

1 Introduction

This chapter is based on the results of the project "Transfer and industrialization of Instant Controlled Pressure Drop (DIC) in the field of fish oil pressing." This project was the basis for a patent, with AZTI-Tecnalia, ABCAR-DIC PROCESS and ULR/LMTAI as the co-owners.

1.1 Fish Oil

1.1.1 Importance of Fish Oil

Unsaturated fatty acids are classified according to the position of the first double bond from the methyl end: between C9 and C10 for ω-9, C6 and C7 for ω-6, and between C3 and C4 for ω-3. The fatty acids of the ω-6 series are linoleic acid (LA; 18:2 ω-6) and arachidonic acid (AA; 20:4 ω-6), with two and four unsaturations, respectively. Fatty acids of the ω-3 series comprise alpha-linolenic acid (ALA; 18:3 ω-3) and the long chain polyunsaturated fatty acids (PUFAs) eicosapentaenoic acid (EPA; 20:5 ω-3) and docosahexaenoic acid (DHA; 22:6 ω-3). A competitive interaction exists whereby fatty acids of the 18:3 ω-3 family suppress the metabolism of the 18:2 ω-6 family, and vice versa.

I. Pérez (✉) • C. Bald • Í. Martínez de Marañón
AZTI-Tecnalia, Astondo bidea g, 609 Eraikina, 48160 Derio, Spain
e-mail: iperez@azti.es

K. Allaf
Laboratory of Engineering Science for Environment (LaSIE FRE 3474) CNRS, 17042 La Rochelle Cedex 01, University of La Rochelle, La Rochelle, France
e-mail: kallaf@univ-lr.fr

T. Allaf and K. Allaf (eds.), *Instant Controlled Pressure Drop (D.I.C.) in Food Processing: From Fundamental to Industrial Applications*, Food Engineering Series, DOI 10.1007/978-1-4614-8669-5_9, © Springer Science+Business Media New York 2014

Fish oil is an excellent source of ω-3, which differs from the oils found in plants in terms of its long chain EPA and DHA, and these have very important physiological roles in growth, reproduction, and brain and eye functions (Murayama et al. 2002). They are also important in preventing certain cancers (Roynette et al. 2004) and decreasing the risk of cardiovascular diseases (Schmidt et al. 2005), depression (Timonen et al. 2004), immune disorders, and others (Bilo and Gans 1990). Furthermore, a large proportion of PUFAs from seed and vegetal oils are of the ω-6 series. Today, the elevated levels of this kind of fatty acid in the human diet mean that fish and fish oil have become the most valuable sources for achieving the ω-6/ω-3 ratio required in our diet.

1.1.2 Quality Guidelines

The most important problem associated with fish oil is rancidity. The high degree of unsaturation of the fatty acids in fish oil means this product is prone to rancidity reactions, which can occur before, during, and after extraction. High temperatures, long cooking times, and metal contaminants favor fast decomposition of the oil.

The hydrolysis of triglycerides liberates fatty acids, which are more prone to oxidation than when esterified with a glycerol molecule. Crude oils with free fatty acid (FFA) contents above 10 % are to be regarded as spoiled and, in most cases, are no longer suitable for use in the food industry as they are difficult to refine. The peroxide value (PV) is related to the primary oxidation products, and the anisidine value (AV) is related to secondary oxidation products. Values of PV and AV higher than 10 mEq/kg and 10, respectively, in the crude oil are typical of a high oxidation level, conferring poor flavor stability on the final product.

The food and pharmaceutical industries are increasingly coming under pressure to meet the requirements of the high-quality standards and to reduce their prices, given the internationalization of the market. They must also comply with constraints imposed to safeguard the environment such as a quantitative and qualitative reduction of waste and energy consumption.

1.2 State-of-the-Art Fish Oil Extraction

Fish is composed of three major fractions: solids (fat-free dry matter), oil, and water. The purpose of fish meal and oil processing is to separate these parts from each other.

1.2.1 Current Methods for Fish Oil Extraction

There are different methods employed in fish oil extraction. We can divide them into two main groups: traditional methods widely employed in factories and new methods with a promising future in the fish oil industry. Numerous papers have been published on this topic but the results have scarcely filtered to the industrial scale.

In the traditional methods, the main steps of the physical processes are: cooking to free the bound water and oil; separation of the solid phase (press cake) and the liquid phase containing water and oil; and recovering oil and solids by successive decantation centrifugation. A final drying and grinding stage can be carried out to recover the remaining solid.

New extraction techniques such as supercritical fluid extraction (SCFE) are more often employed to concentrate ω-3 fatty acids from a mixture of fatty acids (Dunford et al. 1997; Catchpole et al. 2000; Alkio et al. 2000) than to obtain fish oil directly from the structural matrix of fish. The usual approach in these cases is to extract oil from fish using conventional methods, hydrolyze the fatty acids, and esterify them before performing a CO_2-SCFE.

The wet rendering method is the method commonly used at the industrial scale. Most crude oil marketed around the world is fish body oil derived from pressing cooked fish. The process involves first cooking the fish to coagulate the protein and release the bound water and oil. This is followed by pressing to separate the solids from the liquids. The solid material is further dried and then milled to fish meal, and the oil is separated from the liquids by centrifugation (Fig. 1).

Fish meal and fish oil yields can vary depending on the fish species and even from season to season, but typically around 17–22 % fish meal and 5–15 % fish oil will be obtained by processing.

Once the fish is landed, operations to obtain fish oil can be resumed in three main steps:

1. Cooking: the raw fish is passed through a steam-heated continuous cooker, where it is commonly heated to 95–100 °C within 15–20 min. This sterilizes the fish, coagulates the proteins, and disrupts the cell membranes to facilitate the separation of the liquid from the dry matter. Heat is transferred indirectly by a steam-heated surrounding jacket and a heated screw conveyer, which transports the fish material through the cooker. It is known that while fish oil yield is reduced at low temperatures, oxidation is enhanced at high temperatures. Indeed, it has been proved that the total amount of heat has a similar effect; low throughput means a long residence time and consequently unnecessarily high oxidation levels in the oil, which becomes less stable during storage (Aidos et al. 2002, 2003a).
2. Pressing: the purpose of pressing is to squeeze out as much liquid as possible from the solid fish pulp. This is important not only to improve the oil yield and the quality of the meal but also to reduce the moisture content of the press cake as much as possible, thereby reducing the fuel consumption of the driers and increasing their capacity. The most common type of press employed in the fish meal industry is the twin-screw press. The performance of the press depends on the design of the screw and the size of the filter. Press cake moisture content has to be somewhat below 50 %. This objective can be achieved by centrifugation or using “decanters” instead of presses for the first separation of the components of cooked fish.

3. Separation: the press liquor contains a liquid phase (oil plus water) and a solid residue. Both water and solid must be removed from the liquor to obtain the oil. This liquor is usually heated at a temperature of 90–95 °C in order to improve the separation. The separation can be performed in two ways. First, using a two-phase decanter to separate the suspended solids (sludge) followed by a vertical disc centrifuge to separate the water from the oil. Secondly, using a three-phase decanter (tricanter) that simultaneously separates sludge, stick water and oil.

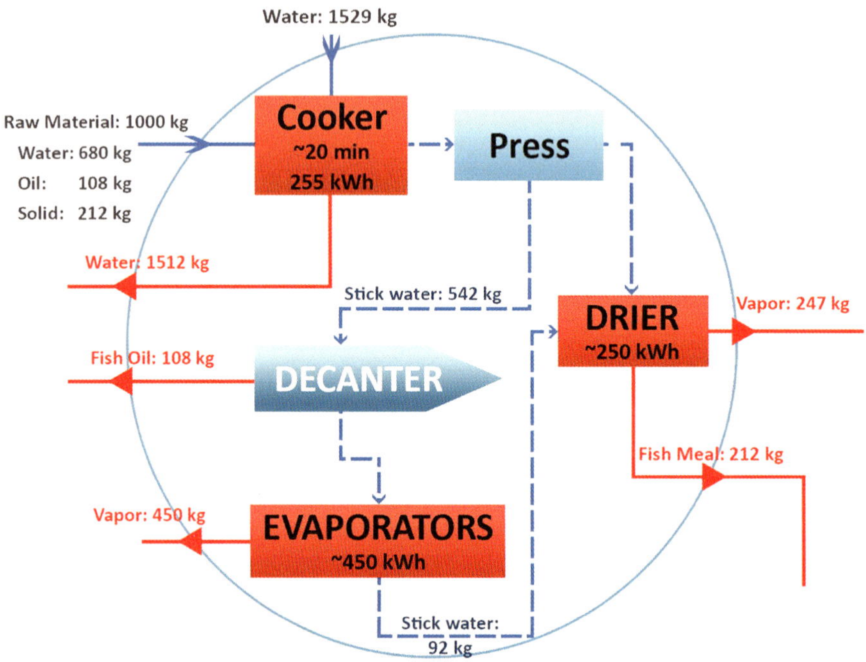

Fig. 1 Traditional wet rendering method for fish oil extraction. Modified from FAO Tech. Pap., 142 (mass and energy balance)

Currently oil obtained from the decanter is not clean enough and must be centrifuged a second time or "polished" to improve its quality. It must be heated to a temperature of 90–95 °C and hot water is added to facilitate the separation in the vertical disc centrifuges. At this stage it is called crude fish oil and its typical chemical and physical properties, defined by the International Fish meal and Fish Oil Organization (IFFO), are listed in Table 1.

Traditional extraction processes have some disadvantages. In the case of the wet rendering method, indirect heating is achieved through a steam-heated jacket. Depending on the nature of the treated material, direct injections of steam and/or longer residence times in the cooker are needed for efficient heat transfer. On the one hand, factors that generally improve oil extraction yield such as temperature and heating time reduce the oxidative stability of extracted oil (Aidos et al. 2003a). Moreover, longer times and higher temperatures mean greater energy consumption. It has been also observed that temperature can even prevent the release of oil since, above a certain level, denatured proteins are closely packed (Chantachum et al. 2000).

On the other hand, it has been found that heat treatment produces oils with higher levels of PUFAs (comprising EPA and DHA) and lower saturated values than extraction by solvents (Aidos et al. 2001, 2003c).

Table 1 Crude fish oil quality guidelines and physical characteristics

Quality guidelines	
Moisture and impurities (%)	Usual basis 0.5 up to 1 % maximum
Free fatty acids (% oleic)	Range 1–7 % but usually 2–5 %
Peroxide value (mEq/kg)	3–20
Anisidine number	4–60
Totox value	10–60
Physical characteristics	
Specific heat (kJ/kg °C)	2.1–2.3
Heat fusion (kJ/kg)	About 225
Caloric value (kJ/kg)	About 39,710
Boiling point (°C)	Greater than 250
Specific gravity (kg/m^3)	
At 15 °C	About 920
At 30 °C	About 910
At 45 °C	About 900
Viscosity (*Pa s*)	
At 20 °C	0.60–0.90
At 50 °C	0.20–0.30
At 90 °C	About 0.10

Source: Bimbo (1998)

Since the products are destined for the food and pharmaceutical sectors, it is necessary to find a compromise between factors that free the oil and those that prevent oxidation and preserve valued molecules such as EPA and DHA. Considering the process as a whole, the aim should be to reduce the cost of the operation while limiting its impact on the environment.

1.2.2 DIC Technology in the Extraction of Fish Oil

In DIC treatment, structural modifications occur as a result of the abrupt autovaporization and these depend on the thermo-hydro-rheological behavior of the material. Under appropriate conditions, the treated product may acquire a uniformly alveolated structure and large exchange surfaces, thereby improving its technological aptitude for subsequent processes.

Swell-drying was also used on seafoods (fish, *Salmo salar* and a *Thunnus albacore*). A direct DIC texturing of raw material followed by airflow drying could preserve the nutritional properties and increase the functionality of the products. The alveolation that is induced results in high rehydration kinetics and a high water holding capacity (WHC), as well as a general preservation of the lipids and proteins, together with the color.

2 Materials and Methods

2.1 DIC Treatment

The DIC equipment consisted of a 92 L vessel and a 20 m^3 vacuum tank (Fig. 2).

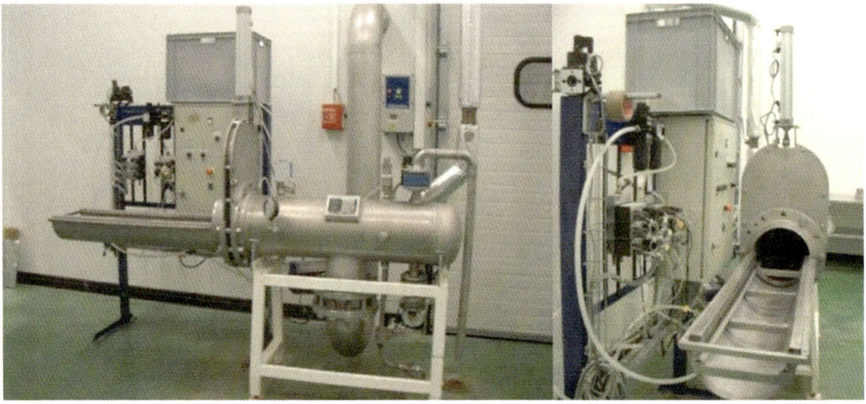

Fig. 2 Pilot plant DIC reactor

The study concerned steam pressure (P) and heating time (t) as quantitative factors and rate of pressure drop (RD), cutting thickness of fillets (Th), and temperature during pressing (PT) as qualitative factors. A 100-fold reduction in the length of the section connecting the reactor to the vacuum tank allowed us to modify the rate of pressure drop (RD) from "High" at 0.16 s to "Low" at 6 s. The objective of this modification was to separate the mechanical effect from the thermal effect of DIC in order to analyze them independently (Table 2). Concerning the temperature of pressing, 80 °C is currently used for the highest yields of fish oil extraction although 30 °C is used for a lower thermal degradation.

Table 2 Experimental points: (Th) slide thickness, (PT) pressing temperature, (*P*) DIC steam pressure, (*t*) DIC heating time, (RD) Rate of pressure drop toward a vacuum (5 kPa)

			Th = 10–20 mm		Th = 3–50 mm	
P (MPa)	*t* (s)	RD	PT = 30 °C	PT = 80 °C	PT = 30 °C	PT = 80 °C
0.2	20	High	Done	Done	Done	–
0.2	20	Low	Done	–	Done	–
0.2	80	High	Done	–	Done	–
0.4	50	High	Done	Done*3	Done	–
0.4	50	Low	Done	–	Done	–
0.4	20	High	Done	–	Done	–
0.6	80	High	Done	Done	Done	–
0.6	80	Low	Done	–	Done	–
Cook 0.1 MPa, 15 min			Done	Done	Done	Done

As already mentioned, two cutting sizes were considered. It is worth noting, for the interpretation of the results, that the initial fat content of the bigger slices (10–20 mm thickness), 33 % db, was superior to that of the smaller thickness (3–5 mm), 25 % db. This was due to the cutting system, which pulled off the skin in thin slices and thus a higher proportion of skin and subcutaneous fat was present in the thick ones.

2.2 Extracting and Recovering Crude Oil

Extraction was performed in a twin-screw press. A co-rotating, intermeshing BC 21 extruder press was used (Clextral, Firminy, France). It was equipped with modular barrels, each 100 mm long, and the total length of the axes was 70 cm. The diameter of the die was slightly superior to that of the screws. The twin screws had segmental 50 mm long screw elements (Ø 25 mm) with variable profiles.

A heated jacket surrounding the barrel sections maintained the experimental conditions at 80 or 30 °C. A K TRON type KCV-KT20 twin-screw feeder fed material into the press inlet port (Fig. 3). The feeding rate was 2.34 ± 0.49 kg fish/h at 67.01 ± 2.93 % wet basis, with a 20 Hz feeder motor frequency and a press screw rotational speed of 130 rpm. Water continuously flowed through the press in order to facilitate the separation of the liquid and solid phases by the 2 mm filter of the press. A GILSON Minipuls Peristaltic pump pumped 1 L water/h into the inlet port of the press.

Fig. 3 View of the pressing system

Two successive centrifugations at 6,640 × *g* for 10 min in a SIGMA 6K 15 centrifuge were performed to recover the oil from the upper phase.

2.3 Assessment of Crude Oil

The fatty acid profile of oil samples and the lipid fraction derived from the flesh of raw, cooked, and *DIC*-treated mackerel were analyzed to determine:

- FFA content or acidity in the oil samples, based on the official UNE 55.011 method. The percentage of FFA was calculated as oleic acid equivalents.
- The peroxide value (PV) of the oil samples was based on the official UNE 55.023 method and expressed in terms of mEq of peroxide per kg of lipid.
- The anisidine value (AV) of the oil samples, which was carried out in accordance with ISO 6885.
- Humidity and volatile matter of the oil samples, determined as the weight loss at 105 °C in a drying oven. Measurements of weight were taken every 30 min until the difference between two consecutive measurements was inferior to 5 %.

3 Results

3.1 Experimental Data

An analysis of variance (ANOVA) was carried out on the data using the statistical program STATGRAPHICS to identify the effect of different experimental factors on several dependent variables.

The dependent variables were Y: oil extraction yield, as a percentage of oil recovered in relation to the pressed matter; LR: residual lipids in the press cake; MR: residual moisture of the press cake; M_I: moisture and impurities in the oil; PV and AV: peroxide value and anisidine value of the oil. The ANOVA of the proposed qualitative variables (cutting size, pressing temperature, and rate of pressure drop during DIC) was used to determine whether significant differences existed between the conditions tested.

3.2 Oil Extraction Yield

An oil yield of 30.78 % db were obtained, which represents about 93 % of the extractable oil and 100 % of the oil extracted after cooking.

Steam pressure, and thus temperature, in the tested range ($P = 0.2$–0.6 MPa and temperature $T \approx 120$–160 °C) did not seem to play an important role, while the heating time negatively affected the extraction of oil. It is worth noting that the faster the drop in pressure, the greater the quantity of extracted oil. Pressing by a system of co-rotating twin screws provided a more consistent and viscous texture.

3.3 Composition of the Press Cake

At first sight, it would appear that lower yields were obtained the higher the temperature in the press barrel. However, the results actually show that the temperature of pressing does not affect the lipid content in the press cake. This may indicate that during pressing the temperature does not modify the extraction behavior but it does influence the quantity of emulsion formed. The difficulty in quantifying this emulsion and its lipid content led us to consider the oil remaining in the solids after pressing as an important index of the pretreatment efficiency of extraction, as oil not present in the residual solids may be recovered. So the goal was to minimize the lipid content.

Considering the conditions that gave the best yield, a low pressing temperature, and a rapid drop in pressure, the optimal values for the DIC parameters were

0.2 MPa of steam pressure and 20 s. The optimized residual oil content in the press cake was 2.39 % db.

The quality of the oil recovered and of the press cake was compared for DIC and cooked samples (Tables 3 and 4).

Table 3 Mean values of the oil quality

	PV (mEq/kg fat)	AV	FFA (% oleic acid)	M_I (% db)	PUFA (g/100 g oil)	EPA (g/100 g oil)	DHA (g/100 g oil)
DIC	23.50	10.44	3.51	0.33	14.54	3.45	5.19
Cook	21.86	11.13	2.31	0.06	14.82	3.57	5.08

PV peroxide value, *AV* anisidine value, *FFA* free fatty acid, *M_I* moisture and impurities, *PUFAs* polyunsaturated fatty acids, *EPA* eicosapentaenoic acid, *DHA* docosahexaenoic acid, *DIC* mean value of DIC treatments, *Cook* mean value of the cooking treatments

Table 4 Mean values for the residual solids, press cake, and composition (% with reference to wet basis)

	Moisture	Fat	Others
DIC	49.40	1.76	48.84
Cook	51.30	3.72	44.98
RM	70.20	8.70	21.10

Others: makes reference to the protein and ash content; DIC, Cook: mean value of the cooking treatments; RM: raw mackerel fillets

The advantages of DIC accrued through the reduction of treatment time, i.e., the oxidative stability of oil and the preservation of the properties of the solids, seemed evident. Aidos et al., working on frozen herring byproducts and comparing these to fresh offal, found higher values for the oxidation state and stability parameters, such as peroxide value (PV), anisidine value (AV), and content of FFAs (Aidos et al. 2003b). With frozen mackerel fillets, we obtained oils with a high level of oxidation of around 22 mEq/kg fat and 11, for PV and AV, respectively (Table 3). The statistical analysis of PV and AV showed a direct ($P < 0.05$) or quadratic ($P < 0.10$) dependency, respectively, on the duration of DIC treatment. However, due to the high oxidations levels, we can only highlight the similarity of both parameters with oil obtained after DIC or traditional cooking.

The FFA content of the oils was within the normal range for a crude oil (2–5 % oleic acid) (Table 3). On the other hand, we noted a higher level of FFAs in DIC oil in general. It would be interesting to determine if decompression plays a role in the liberation of these fatty acids.

The results concerning the concentration in PUFAs, especially EPA and DHA, are presented in Table 4. Raw fish contained 11.77, 2.32, and 5.94 g/100 g oil of PUFAs, EPA, and DHA, respectively. Contents of around 15 % PUFAs in oils extracted by pressing confirmed that heat treatment produces oils with higher levels of PUFAs than extraction using solvents (Aidos et al. 2001). In consequence, it is necessary to limit the discussion to a comparison of oils obtained by both heating treatments. Similar mean values have been found for oil obtained after DIC treatment and after cooking.

Considering the proportion of the phases produced, it is noticeable that the values of the solid and water phases were practically inversed. The aqueous phase produced in the process incorporating DIC technology was superior to that obtained by the cooking process, which eliminated more water from the product matrix and thus gave dryer solids. This could be explained by the results obtained by Guyomard (1994) in his thesis concerning twin-screw pressing. He found that the mechanical squeeze effect, intensified by a particular configuration of the screws, allowed a better release of liquids while simultaneously reducing the yield of extracted oil. In the present case, we attributed this effect to DIC treatment, as it was not produced in the process using traditional heating. However, this effect actually does seem to favor the formation of emulsion, probably by a more intense mixing of the three phases composing the liquid that is released.

A slower flow rate of cooked fish fed into the press meant a longer residence time of treated fish in the extruder. As it is known that decreasing the feeding rate favors extraction (Aidos et al. 2003a), a reduction of the rotating speed of the screws should improve the oil extraction yield.

3.3.1 Conditions of Fish Pretreatment by DIC for the Extraction of Oil by Pressing

Figure 4 represents the different stages of the process and the general mass balance at the principal phases of the process: thermal/thermomechanical treatment and extraction–separation.

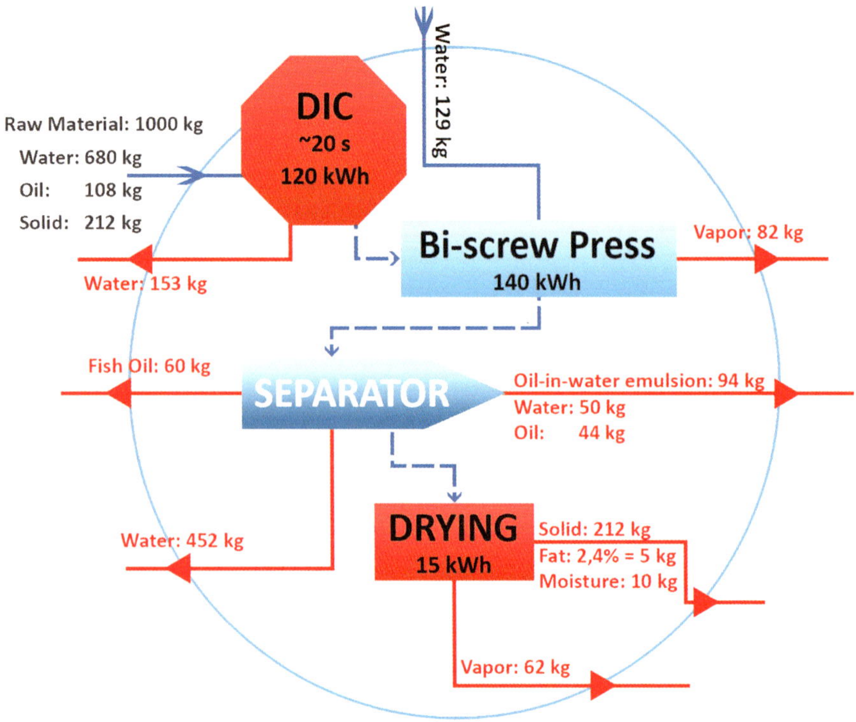

Fig. 4 DIC method for fish oil extraction (mass and energy balance)

Taking into account the experimental results and optimal conditions established by the models, as well as the diversity of the raw materials available for the production of fish oil, the process for fish oil extraction by mechanical pressing has to include a thermomechanical pretreatment by DIC in the following conditions (Table 5):

Table 5 Treatment conditions proposed for DIC thermomechanical pretreatment of fish in the oil extraction process by twin-screw pressing

Pressure	0.20 or 0.60 MPa
Time of thermal treatment	20 s
Mincing thickness	Depending on the raw material
Pressure drop rate	The highest (between 1 and 4 MPa/s)
Thickness of treated layer	12 cm

With our application, we observed that a homogeneous treatment does not favor subsequent pressing of the fish matter. The necessity to create a certain difference in texture throughout the treated matter allows the treatment of thick layers of between 10 and 17 cm. We propose a thickness of around 12 cm to obtain a good technological aptitude of the matter for pressing.

4 Conclusions

The process of oil extraction includes several phases, of which heating is merely a pre-stage of the actual extraction step. It is not possible to attribute the success of the whole process to any single phase, which is why we cannot talk about the improvement of the extraction by DIC but rather about the beneficial modifications to the product structure that improve the extraction by pressing.

The general objective of heating is to break down bonds retaining oil in the structural matrix by the coagulation of proteins. However, conditions providing greater degradation are not necessarily appropriate for a subsequent extraction by pressing as the flow of already liberated liquid can be hindered.

In this regard, the creation of an internal gradient seems to be indispensable for the correct pressing of the treated matter. Either a specific distribution of the moisture content within the material or a nonhomogeneous treatment with regard to the whole treated mass confers a specific and favorable rheology to the material to be pressed.

By using co-rotating twin screws as a pressing system, more than 70 % of the theoretically extractable oil can be recovered. In this regard, pressing with Clextral's BC21 co-rotating twin-screw press has allowed us to draw conclusions about the efficacy of DIC as a pretreatment for fish oil extraction.

The instantaneity of decompression in DIC has a mechanical effect that complements the thermal action, with a specific effect on the structure of the matter. The fact that there is no significant effect of steam pressure on the oil yield indicates that increasing the treatment temperature does not produce the equivalent effect on the product. This enables a reduction in thermal treatment time, compared to traditional cooking, from 15 min to 20 s, with equivalent or even superior extraction yields and a consequent preservation of the product quality.

It is possible, using thermomechanical DIC treatment and co-rotating pressing, to obtain a press cake with a fat content of up to 2 % db, which is nearly 50 % of that obtained after traditional cooking in the same conditions of pressing. This result introduces important advantages in relation to the production of low-fat fish meal.

References

Aidos I, Van Der Padt A, Boom RM, Luten JB (2001) Upgrading of maatjes herring byproducts: production of crude fish oil. J Agric Food Chem 49(8):3697–3704. doi.10.1021/jf001513s

Aidos I, Masbernat-Martinez S, Luten JB, Boom RM, Van Der Padt A (2002) Composition and stability of herring oil recovered from sorted byproducts as compared to oil from mixed byproducts. J Agric Food Chem 50(10):2818–2824. doi:10.1021/jf011318t

Aidos I, Kreb N, Boonman M, Luten JB, Boom RM, Van Der Padt A (2003a) Influence of production process parameters on fish oil quality in a pilot plant. J Food Sci 68(2):581–586. doi:10.1111/j.1365-2621.2003.tb05714.x

Aidos I, Schelvis-Smit R, Veldman M, Luten JB, Van Der Padt A, Boom RM (2003b) Chemical and sensory evaluation of crude oil extracted from herring byproducts from different processing operations. J Agric Food Chem 51(7):1897–1903. doi:10.1021/jf020684p

Aidos I, Van Der Padt A, Boom RM, Luten JB (2003c) Quality of crude fish oil extracted from herring byproducts of varying states of freshness. J Food Sci 68(2):458–465. doi:10.1111/j.1365-2621.2003.tb05694.x

Alkio M, Gonzalez C, Jäntti M, Aaltonen O (2000) Purification of polyunsaturated fatty acid esters from tuna oil with supercritical fluid chromatography. J Am Oil Chem Soc 77(3):315–321. doi:10.1007/s11746-000-0051-3

Bimbo AP (1998) Guidelines for characterizing food-grade fish oil. Inform 9(5):473–481

Bilo HJG, Gans ROB (1990) Fish oil: a panacea? Biomed Pharmacother 44(3):169–174, http://dx.doi.org/10.1016/0753-3322(90)90005-T

Catchpole OJ, Grey JB, Noermark KA (2000) Fractionation of fish oils using supercritical CO2 and CO2+ethanol mixtures. J Supercrit Fluids 19(1):25–37, http://dx.doi.org/10.1016/S0896-8446(00)00075-9

Chantachum S, Benjakul S, Sriwirat N (2000) Separation and quality of fish oil from precooked and non-precooked tuna heads. Food Chem 69(3):289–294, http://dx.doi.org/10.1016/S0308-8146(99)00266-6

Dunford NT, Temelli F, LeBlanc E (1997) Supercritical CO2 extraction of oil and residual proteins from Atlantic mackerel (*Scomber scombrus*) as affected by moisture content. J Food Sci 62(2):289–294. doi:10.1111/j.1365-2621.1997.tb03987.x

Guyomard P (1994) Etude de faisabilité d'un extrudeur bivis en pressage-extrusion de graines oléoprotéagineuses = Faisability study of a co-extruder in expression-extrusion of oleaginous seeds. Université de Technology de Compiègne, Compiègne, France

Murayama K, Yoneya S, Miyauchi O, Adachi-Usami E, Nishikawa M (2002) Fish oil (polyunsaturated fatty acid) prevents ischemic-induced injury in the mammalian retina. Exp Eye Res 74 (6):671–676, http://dx.doi.org/10.1006/exer.2002.1151

Roynette CE, Calder PC, Dupertuis YM, Pichard C (2004) n-3 Polyunsaturated fatty acids and colon cancer prevention. Clin Nutr 23(2):139–151, http://dx.doi.org/10.1016/j.clnu.2003.07.005

Schmidt EB, Arnesen H, de Caterina R, Rasmussen LH, Kristensen SD (2005) Marine n-3 polyunsaturated fatty acids and coronary heart disease. Part I. Background, epidemiology, animal data, effects on risk factors and safety. Thromb Res 115(3):163–170

Timonen M, Horrobin D, Jokelainen J, Laitinen J, Herva A, Räsänen P (2004) Fish consumption and depression: the Northern Finland 1966 birth cohort study. J Affect Disord 82(3):447–452, http://dx.doi.org/10.1016/j.jad.2004.02.002

Closing Remarks: Instantaneity and Autovaporization

Karim Allaf

Theoretical studies on the DIC process are based on instantaneity thermodynamics (vs. quasi-static thermodynamics). Instantaneous transformations such as instant autovaporization are characterized by an anisotropic (vs. isotropic) translation movement of particles with a transitory (vs. permanent) process. Indeed, by allowing the normally isotropic three-dimensional translational fluctuation movement of particles to occur in only two dimensions, true thermodynamic temperature asymptotically reaches a minimal level $\theta_m = 2/3\ \theta_i$. After this very short transitory period, a thermodynamic equilibrium can be reached through quasi-static transformation.

For instant transformations, zero change in the total entropy in the universe can be considered as an asymptotic approach to ideal irreversible instant transformations. Allaf has suggested the impossibility of generating a flow of infinite total entropy in the universe:

Finite amount of total entropy flow in the universe:

$$\left.\frac{\partial S}{\partial t}\right]_{t\to 0} = \text{finite quantity} \tag{1}$$

which means that the amount of the total entropy in the universe during instantaneous transformation should asymptotically be null:

$$\Delta S]_{t\to 0} = 0 \tag{2}$$

Thus, after a very short isentropic "flash autovaporization," the system normally crosses the quasi-static (three-dimensional translational fluctuation) temperature to

K. Allaf
Laboratory of Engineering Science for Environment (LaSIE FRE 3474) CNRS,
17042 La Rochelle Cedex 01, University of La Rochelle, La Rochelle, France
e-mail: kallaf@univ-lr.fr

T. Allaf and K. Allaf (eds.), *Instant Controlled Pressure Drop (D.I.C.) in Food Processing: From Fundamental to Industrial Applications*, Food Engineering Series,
DOI 10.1007/978-1-4614-8669-5_10,

reach a level lower than the equilibrium. The fluctuation velocity of molecules should be established as an isotropic, permanent, and homogeneous distribution obtained after this quasi-static part of the transformation; the amount of the total entropy in the universe then normally increases. Experiments confirmed these theoretical analyses that are specific to the instantaneity phenomena. They reveal that instant autovaporization is a very short, anisotropic, and transitory operation. It is followed by a quasi-static type transformation, which is an isotropic, permanent, and homogeneous translational fluctuation movement of particles.

In all the studies conducted by Allaf and collaborators, the rate of the pressure drop was of particular interest for DIC treatment. Indeed, the results obtained when DIC was used to treat organic products revealed that the autovaporization ratio, the expansion ratio, the specific surface area, and the cooling rate of the products were much higher than in non-instantaneous (quasi-static) decompression.

Studies on dehydration by multi-flash drying (MFD) showed that the drying rate was substantially faster than with quasi-static decompressions. The DIC extraction of volatile compounds (e.g., essential oils) has demonstrated that it is closely related to instant autovaporization. In DIC spraying conducted on pumpable materials, it was found that the closer the spray nozzle, the greater the amount of water removed.

Swell-drying, coupling conventional drying processes with expansion by DIC, provides a very effective intensification of the operation. It results in lower energy consumption, faster kinetics, and a higher functional, nutritional, and health quality of the end product.

DIC used as an instant autovaporization usually allows essential oil extraction and deodorization in minutes, whereas hours or even days are needed for conventional processes. This treatment simultaneously involves an expansion, which increases the porosity of the material. Whatever the plant organ or conventional solvent (including supercritical CO_2), the extraction operations as well as new concept processes are then much better adapted for the extraction of plant and nonvolatile active molecules.

These results are of great importance and have enormous technological implications. They have been confirmed in specific and interesting studies in various areas with different industrial applications.

Index

T. Allaf and K. Allaf (eds.), *Instant Controlled Pressure Drop (D.I.C.) in Food Processing: From Fundamental to Industrial Applications*, Food Engineering Series, DOI 10.1007/978-1-4614-8669-5, © Springer Science+Business Media New York 2014

Printed by Printforce, the Netherlands